Worm and Spiral Gearing

Gear Design and Engineering

BY

Frederick A. Halsey

Reprinted From
"American Machinist"

ISBN 1-929148-31-3

Wexford College Press 2003

PREFACE.

JUSTIFICATION for the republication of the contents of this book is found in the still prevalent opinion among designers of machinery that worm gearing is necessarily short lived and of low efficiency, and in the fact that the methods of laying out spiral gearing are not as widely understood as the merit and convenience of that form of gearing make desirable.

The theory of worm gearing is well fortified by the collection of facts from experience given herein, and it points out clearly the procedure to be followed in order to insure durability and efficiency. Both analytical and graphical methods of laying out spiral gearing are given, and these, it is believed, will meet the needs and tastes of all.

It should be mentioned that with spur gears, when stock cutters are used, the diameters and center distances must be such as go with whole numbers of teeth. The property of spiral gears by which any center distance may be accommodated is therefore one which is not shared by spur gears. Should those sections of this book which relate to unchanged center distances be omitted, its remaining solutions would be as complete as is possible with spur gears.

As uncertainty has arisen in the minds of some readers, it should also be mentioned that the solutions here given relate exclusively to gears having their shafts at right angles.

CONTENTS

PART I

WORM GEARING

PART II

SPIRAL GEARING

Wexford College Press

6

ILLUSTRATIONS.

The illustrations are too large to go in the text where they belong, and, with the exceptions noted below, will be found, folded, at the end of the book.

PART I.

WORM GEARING.

In view of the good results now being obtained with worm gearing, the old prejudice against that form of gearing, on account of its supposed low efficiency and short life, is dying out. These good results are the outcome of the application of principles which are by no means a late discovery, and it is expected that what follows will contain much that to some readers is not new. At the same time it is an undoubted fact that the best practice with worms is understood by but few, relatively speaking, and the corroboration of the theory by examples from practice which follow, is believed to be new. No better illustration of the fact that good practice with worm gearing is not

yet widely understood could be given
than the statement in a recent and excel-
lent work on gearing that "the diameter
of the worm is commonly made equal to
four or five times the circular pitch," the
fact being that such proportions are distinctly bad if the worm is to do hard
work.

It should be stated at the beginning
that while what follows is not offered as
a presentation of all the data necessary
for assured success with worms under all
conditions, it is hoped to make the general conditions of successful practice
plain, and to present the "state of the
art" as it exists to-day.

The essential change in practice which
has improved the results obtained with
worm gearing has been an increase in the
pitch angle over what was formerly considered proper. There is no doubt what-
ever that this change has increased the
efficiency of the gear, and, what is of
more importance, has reduced the tendency to heat and rapid wear. This is

not only a fact, but it is a sound conclusion from theoretical considerations,
which might have been predicted under
proper examination.

THEORY OF WORM EFFICIENCY.

The reason why an increase of pitch,
other things being equal, or in other
words, an increase of the angle of the
thread, gives these results, will be understood from Fig. 1. If a b be the axis of
the worm and c d a line representing a
thread, against which a tooth of the wheel
bears, it will be seen that if the tooth
bears upon the thread by a pressure P,
that pressure may be resolved into two
components, one of which, e f, is perpendicular, while the other, e g, is parallel to
the thread surface. The perpendicular
component produces friction between the
tooth and the thread. The useful work
done during a revolution of the thread is
the product of the load P and the pitch of
the worm, while the work lost in friction
is the product of the perpendicular pres-

sure *e f*, the coefficient of friction and the distance traversed in a revolution, which is the length of one turn of the thread. Now, if the angle of the thread be doubled, as indicated, the load *P* remain-

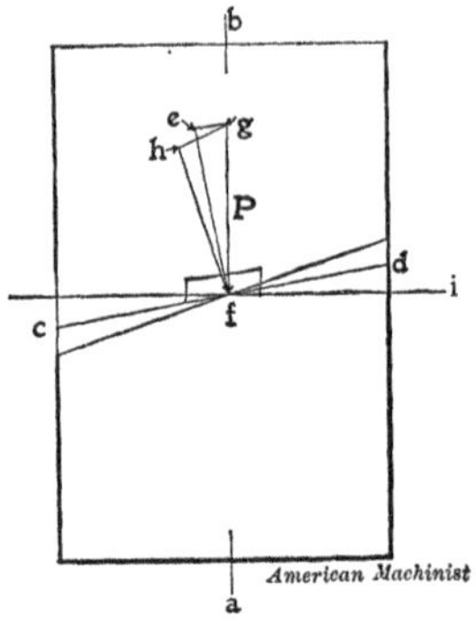

Fig. 1. THE PRINCIPLE OF WORM EFFICIENCY.

ing the same, the new perpendicular component *f h* of *P* will be slightly reduced from the old value *e f*, while the length of a turn of the thread will be slightly increased. Consequently their product and the lost work of friction per revolution

will not be much changed. The useful
work per revolution will, however, be doubled, because, the pitch being doubled,
the distance traveled by P in one revolution will be doubled. For a given amount
of useful work the amount of work lost is
therefore reduced by the increase in the
thread angle, and, since the tendency to
heat and wear is the immediate result of
the lost work, it follows that that tendency
is reduced. For small angles of thread
the change is very rapid, and continues,
though in diminishing degree, until the
angle reaches a value not far from 45 degrees, when the conditions change and
the lost work increases faster than the
useful work, an increase of the angle of
the thread beyond that point reducing
the efficiency.

This general consideration of the subject shows the principles at the bottom
of successful worm design, but a more
exact examination is desirable. According to Professor Barr the efficiency of a

worm gear, the friction of the step being neglected, is :

$$e = \frac{\tan \alpha \, (\, 1 - f \tan \alpha)}{\tan \alpha + f}$$

in which

$e =$ efficiency,

$\alpha =$ angle of thread, being the angle $d\,f\,i$ of Fig. 1,

$f =$ coefficient of friction.

To study the effect of the step, a convenient assumption is that the mean friction radius of the step is equal to that of the worm. This assumption would only be realized in cases where the step is a collar bearing outside the worm shaft, and the preceding and following formulas therefore represent extreme cases, one of a frictionless step, which would be approximated by a ball bearing, and the other of a step having about the extreme friction to be met with. Most actual cases would therefore fall between the two. Again, according to Professor

Barr, the efficiency of a worm and step
on the above assumption is : *

$$e = \frac{\tan \alpha (1 - f \tan \alpha)}{\tan \alpha + 2f} \text{ (approximately)}$$

Notation as before.

These formulas give no clear indica-
tion of the manner in which the efficiency
varies with the angle, and the diagram,
Fig. 2, has been constructed to show this
to the eye. The scale at the bottom
gives the angles of the thread from 0 to
90 degrees, while the vertical scale gives
the calculated efficiencies, the values of
which have been obtained from the equations and plotted on the diagram. The
upper curve is from the first equation,
and gives the efficiencies of the worm
thread only; while the lower curve, from
the second equation, gives the combined
efficiency of the worm and step. In the

---

* In Professor Barr's formulas it is assumed that the
worm thread is square in section. Thread profiles in com-
mon use affect the results but little.

calculations for the diagram it is necessary to assume a value for f, and this has been taken at .05, which is probably a fair mean value. The experiments made by Mr. Wilfred Lewis for Wm. Sellers & Co. showed an increase of efficiency with the speed. The present diagram may be considered as confined to a single speed, and at the same time is not to be understood as showing the exact efficiency to be expected from worms, but rather to exhibit to the eye the general law connecting the angle of the thread with the efficiency.

The curves will be seen to rise to a maximum and then to drop. The exact values of the angle of thread to give maximum efficiency may be easily found by the methods of the calculus, the results being :

For worm thread alone the efficiency is at a maximum when

$$\tan \alpha = \sqrt{1 + f^2} - f.$$

Substituting the value of f (.05) used in calculating the diagram, this becomes tan α for maximum efficiency $= .9512$,

and by referring to a table of natural tangents we find that

α for maximum efficiency $= 43° \; 34$.

Similarly for the worm and step the result is

tan α for maximum efficiency $=$

$\sqrt{2 + 4 f^2} - 2 \; f$, which for $f = .05$ $= 1.318$,

and a table of tangents tells us again that

α for maximum efficiency $= 52° \; 49'$.

Of more importance than the angle of maximum efficiency is the general character of the curves, of which the most pronounced peculiarity is the extreme flatness, showing that for a wide range of angles the efficiency varies but little. Thus, for the upper curve there is scarcely any choice between 30 and 60 degrees of angle, and but little drop at 20 degrees.

At first sight the lower curve might be

thought the most useful of the two, as it
includes the effect of the step, but a little
consideration will show that this is not
the case. For most cases in which worms
are used the efficiency of the transmis-
sion, as such, is of very little account.
What the designer concerns himself with
is the question of durability and satisfactory working, and the results to be expected in this respect are best shown by
the upper curve, in which high efficiency
means a durable worm.   Throughout
this discussion, in fact, the chief significance of efficiency lies in the fact that
low efficiency means rapid wear and vice
versa.

EXPERIMENTAL CORROBORATION OF THE THEORY.

The experiments of Wm. Sellers & Co.,
before referred to, go far to confirm the
soundness of the above views.   From the
present standpoint it is unfortunate that
those experiments did not cover a wider
range of worm thread angles—those act-

ually used being 5 degrees, 7 degrees,
and 10 degrees. Other experiments were,
however, made on spiral pinions of higher angles, spiral pinions being understood
by Mr. Lewis to mean those pinions
having the mating gear a true spur, the
pinion shaft being at a suitable angle
with the gear shaft to bring the pinion
in proper mesh—a construction which
is exemplified in the well-known Sellers
planer drive. Mr. Lewis gives a formula
by which the efficiencies of worms can be
calculated from those for spiral pinions,
and in the absence of direct experiments
on worms of high angles, his results for
spiral pinions have been modified by this
formula to read for worms. The results
for the two forms of gearing differ by
less than five per cent. for the extreme
case of his experiments. To compare the
results obtained by Mr. Lewis with Professor Barr's formula, a speed has been
selected from the experiments giving the
nearest coefficient of friction to that used
in obtaining the curves of Fig. 2. The

results have been plotted in Fig. 2, where
they appear as small crosses, and will be
seen to have a very satisfactory agree-
ment with the lower curve, with which
they should be compared, as the steps of
the worms used by Mr. Lewis were of
the usual pattern without balls.

The variation of the coefficient of friction with the speed lends an interest to
Fig. 3, which is a series of curves obtained from the results published by Mr.
Lewis in the same manner as the crosses
of Fig 2, the curve for 20 feet velocity
being in fact the same as that appearing
as crosses in Fig. 2. The other curves
of Fig. 3 are obtained from those of Mr.
Lewis, and cover a range of velocities
from 3 to 200 feet per minute at the pitch
line, as noted at the right. In this diagram the results obtained by Mr. Lewis
on worms are plotted direct, but the experiments on spiral pinions have been
modified as explained above. Inspection
of the curves shows that while there is a
progressive increase of efficiency with the

speed, there is, nevertheless, not much probability, or indeed room, for further improvement beyond the speed of 200 feet per minute. It will furthermore be seen that the efficiency drops off much less for low angles of thread at high speeds than at low.

In interpreting this diagram, it should be remembered that the durability of a worm depends upon the amount of power lost in wear, and not upon the percentage so lost. The ability of a given worm to absorb and carry off the heat due to friction is fixed, and does not vary with the speed. That is, a given worm running at 100 revolutions under a given pressure can carry off as much friction heat as the same worm at 200 revolutions, while it, under the same pressure would transmit but one-half the power in the former case that it would in the latter. In other words, the percentage of lost work might be twice as much at the lower speed as at the higher without increasing the tendency to heat.

The increase of efficiency with the speed is a valuable property of worms, and enables them to do much more work than they otherwise would. Thus the 20 degree worm at 20 feet per minute lost $21\frac{1}{2}$ per cent. of the work in friction. Increasing the speed to 40 feet doubled the work applied, and, had the efficiency remained constant, would have doubled the friction heat to be dissipated. In point of fact, this increase of speed diminished the percentage of loss to 17, and the amount of loss and heat, instead of being doubled, was only increased in the ratio of 160 to 100. It is plain from the diagram, however, that this action does not continue much beyond a velocity of 200 feet per minute, beyond which the amount of loss must be more nearly proportional to the speed, and this doubtless has some connection with the fact observed by Mr. Lewis that 300 feet per minute is the limit of speed when the gears are loaded to their working strength, and that the best conditions are obtained at

about 200 feet per minute. It is proper
to add, however, that in the cases from
practice given later there are three which
have been made repeatedly, and which
are conspicuously successful, in which
the velocity exceeds 600 feet, and one in
which it exceeds 800 feet. No doubt, in
all such cases, if the pressure on the teeth
could be known it would be found to be
light.

It will be seen that an increase of speed
for any worm under constant pressure
leads to an increase of friction work, and
the limit is reached when the worm is no
longer able to carry off the heat gener-
ated fast enough to prevent undue rise
in temperature. Furthermore, this limiting speed depends upon the pressure,
it being higher for low pressures than
for high. A worm having an angle
which might be successful at low speed
may fail at high speed; but it would seem
that any worm which is successful at
high speed should also be successful at
low, which is in accordance with mechanical instinct.

There are, it will be observed, two
methods of increasing the pitch angle.
The diameter may be kept constant and
the pitch be increased, or the pitch may
be kept constant and the diameter be reduced. From a mathematical standpoint,
these two methods are identical; that is,
at a given pitch line velocity a worm of a
given angle should have the same efficiency, regardless of the diameter; but
in a mechanical sense the methods are
not identical. The worm of the larger
diameter would naturally have a gear of
wider face and the pair, having greater
area of tooth surface in contact would
carry a larger load.

EXAMPLES FROM PRACTICE.

It is impossible to say who was the
first to recognize the significance of the
pitch angle as a factor in the satisfactory
performance of worm gearing, but it may
be mentioned as a matter of interest that
the exhibit of the Hewes & Phillips Iron
Works at the Newark Industrial Exhibi-

tion of 1873 included several worm-
driven planers, in which the worms were
double threaded and had a pitch angle of
15° 15′, a pitch diameter of $3\frac{1}{2}$ inches, a
pitch of 3 inches, and·a speed, cutting, of
256 and backing of 640 r. p. m., which
give pitch line velocities of 237 and 590
feet. This worm was successful, and was
many times repeated; but later on Hewes
& Phillips were struck by the high belt
speed idea, and in order to increase the
belt speed they changed the worm to 6.16
p. d., $1\frac{3}{4}$ inches pitch, single thread;
speed, cutting, 446, and backing 1,110 r.
p. m., giving a pitch angle of 5° 15′ and
pitch line velocities of 720 and 1780 feet.
This worm was a failure, and was soon
changed to 6.16 p. d., $3\frac{1}{2}$ inches pitch,
double thread; speed, cutting, 281, and
backing 700 r. p. m., giving an angle of
10° 15′ and pitch line velocities of 452 and
1,130 feet. This worm did better than
the last, but not so well as the first. By
this time the lesson was learned, and
Hewes & Phillips set out to use a worm

of 30 degrees pitch angle. Structural considerations, however, prevented the use of so high an angle and they compromised on 20 degrees, the final worm resulting from this experience having a pitch diameter of 2.63 inches, with 3 inches pitch, quadruple thread, the speed cutting being 300 and backing 700 r. p. m., giving pitch line velocities of 205 and 480 feet, and this remained the standard angle as long as these planers were manufactured. The writer has seen one of these 20 degree worm gears, opened up after twelve years' use, and the wear disclosed was very slight—no shoulder being in existence. As a result of the experience outlined above, this house adopted the standard practice of making the worms as small as possible in diameter, and giving the threads in all cases a pitch angle of 20 degrees. The form of tooth used was the epicycloidal, while the materials used were hard cast-iron for the gear and case-hardened open-hearth steel for the worms.

These Hewes & Phillips worms are plotted in Fig. 3 as crosses 1, 2, 3, 4, of which 1 is the 15° 15′, 2 the 5° 15′, 3 the 10° 15′, and 4 the 20°, the first and last being successes, and the second and third failures.

In plotting these worms, and all others having pitch line velocities above 200 feet, the crosses are placed near and above the 200 feet curve. It is unfortunate that we have no curves for higher speeds, but Mr. Lewis recommends the use of the 200 feet line for all higher speeds. Leaders connecting different crosses indicate the same worm at different speeds in all cases. The letters s and f on the diagram mean success or failure in all cases.

Fig. 4 is a drawing of worm 3·(failure) and Fig. 5 shows worm 4 (success), and no more instructive pair of drawings could be imagined than these. The pitches are not far different, and what difference there is is in favor of the larger worm. The duty is the same, the gears are of about the same diameter, and

the revolutions per minute are nearly the
same. The essential change is in the increase of the pitch angle by a reduction
of the diameter, and this changed failure
to success.

The Newton Machine Tool Works use
worm gearing in many of their machines,
notably their cold saw cutting-off machines. In the earlier machines of this
class the worm had a pitch diameter of
$2\frac{7}{8}$ inches, with a pitch of 1 inch, single
thread, the revolutions per minute being
765. These figures give a pitch angle of
6° 20', and a pitch line velocity of 572
feet. This machine could be operated,
but not with satisfaction on account of
the heating and short life of the worm.
The worm was then increased in pitch by
making it double threaded, giving a pitch
angle of 12° 30', the speed being reduced
to 500 revolutions per minute, giving a
pitch line velocity of 375 feet. The
change proved to be a great improvement, heavier work than was before possible being done after the change without

distress or difficulty, and this worm has
since been applied to a large number of
machines with entire success. A still
later worm used on these machines has a
pitch diameter of $3\frac{7}{8}$ inches and a pitch
of 4 inches. triple threads, giving a pitch
angle of 18° 15', and this is found to be a
still further improvement. This last
worm is used on a wide variety of machines and at a variety of speeds from 40
to 680 r. p. m., giving pitch line velocities
of from 40 to 685 feet, and with uni-
formly good results. In many cases it is
used without an oil cellar, though for
comparatively light work. The form of
thread used is the involute, and the material is hardened steel for the worm and
bronze for the wheel. These Newton
worms appear in Fig. 3 as 5, 6, 7, of
which 5 is nearly a failure, while 6 and 7
are entirely successful. The second Newton worm—the one appearing in Fig. 3
as 6—is shown in Fig. 6.

Another habitual user of worms is John
Bertram & Sons, of Dundas, Ontario, Can-

ada, who employ them in all their planers, and use largely a worm of 3.18 inches pitch diameter, 4 inches pitch, quadruple threads, the speed, cutting, being 186 and reversing 744 r. p. m. These figures give a pitch angle of 22 degrees, and pitch line velocities of 155 and 620 feet. This worm appears in Fig. 3 as 8, the vertical position for the higher speed being again uncertain. These worms are highly successful, as the writer knows from repeated observation. Both worm and wheel are of cast-iron, the thread being Brown & Sharpe standard. The Bertram worm is shown in Fig. 7. In reading this drawing it should be remembered that the conventional representation of a worm, with the threads shown by straight lines, shows a larger apparent pitch angle than the true one, as shown by a true projection.

Another case of failure was a worm drive applied to a large boring machine, the worm being 12 inches pitch diameter, 8 inches pitch, quadruple thread, speed

80 r. p. m. and above, worm of forged
steel, wheel of bronze, oil cellar lubrica-
tion. These figures give a pitch angle
of 12 degrees and a pitch line velocity of
250 feet. This worm is located on Fig. 3
as 9.

Still other cases of change from failure
to success are supplied by Mr. Jas. Christie, of the Pencoyd Iron Works. The
first of these relates to a boring machine,
which was, by the makers, supplied with
a worm drive having a worm of $5\frac{1}{4}$ inches
pitch diameter, $1\frac{1}{2}$ inches pitch, single
thread, steel worm and cast-iron wheel,
average speed 150 r. p. m. These figures
give a pitch angle of 5 degrees and a pitch
line velocity of 215 feet. This was a failure, but was successfully replaced by a
worm of $4\frac{7}{8}$ inches pitch diameter, $2\frac{1}{2}$
inches pitch, and the same number of
revolutions, which figures give a pitch
angle of $9°$ $15'$ and a pitch line speed of
190 feet. These two worms appear as 10
and 11. This successful worm lies in the
region of unsuccessful ones, but the in-

fluence of the increased lead angle is unmistakable. The fact of its success is
probably due to the pressure on the teeth
being well below the working strength,
or to the speed being moderate, or both.

The second case, of which the data
were supplied by Mr. Christie, relates to
two heavy milling machines, in which the
cutter spindles were driven by worms 6
inches pitch diameter by $1\frac{1}{2}$ inches pitch,
single thread. It was found that the cutters could be run much faster than was
originally contemplated, and the worms
were consequently speeded up to about
500 r. p. m. In these machines cast-iron
worm wheels were speedily destroyed,
while hardened steel worms and bronze
wheels would last about a year. Later
two more machines were built having steel
worms and bronze wheels, the worms be-
ing $4\frac{1}{2}$ inches pitch diameter by 5 inches
pitch, quadruple threads, speed 280
r. p. m. These worms have been in use six
years, and are described as being "good
as new." The data given for the first

worm give a pitch angle of 4° 30′ and a pitch line velocity of 785 feet. It appears in Fig. 3 as 12. The pitch angle of the second worm is 19° 30′, and its pitch line velocity 328 feet. It appears in Fig. 3 as 13.

Mr. Christie has made many successful changes, of which these are typical, and he now uses worms with great freedom and success. His general conclusion is that good worms begin with those having the pitch about equal to the diameter, giving a pitch angle of 17° 15′.

Another equally striking case of success accompanying an increase of the pitch angle is supplied by Mr. W. P. Hunt, of Moline, Ill., who says :

"In building a special double-spindle lathe I wished to use a worm drive, and having a single-thread ¾-inch pitch hob, 2¾-inch outside diameter, I decided to work to that, and made my gear with twenty-six teeth, giving a speed reduction of 26 to 1. The worm was to ruu at 460 revolutions per minute, but upon starting

the machine I found it impossible to keep
the worm and gear cool, and the belts
would not pull the cut.

" Accordingly I decided to make a new
worm and hob having the same outside
diameter as the one first tried, but with
double thread and 2-inch pitch, and a
new gear having forty-eight teeth, giving
me a speed reduction of 24 to 1, or less
than at first.

" Upon starting the machine with the
new worm and gear, not only did it run
perfectly cool, but the belts have ample
power. We use graphite and oil on the
worm, and it is not enclosed."

Mr. Hunt does not give the pitch diameter of his worms, but assuming the
threads to have been in accordance with
the Acme standard, the pitch diameters
are $2\frac{3}{8}$ and $2\frac{1}{4}$ inches respectively, the
thread angles being $5°\ 44'$ and $15°\ 48'$,
and the pitch line speeds 286 and 271
feet per minute. Mr. Hunt's worms are
plotted in Fig. 3 as 17 and 18.

Three other cases of successful worms

under heavy duty are found in milling machines which have been repeated many times. The first two worms would ordinarily be described as spiral gears. The shafts are at right angles and the action is that of worms, and they are properly included here.

The first of these, which appears as 14 in the diagram, has a pitch diameter of $2\frac{1}{4}$ inches, a pitch angle of 45°, and a speed varying between 180 and 945 r. p. m., giving pitch line velocities of 106 to 555 feet per minute. Both gears are of cast-iron. The second, 15 in the diagram, is of the same style, and has the same pitch diameter, with speeds varying between 90 and 472 r. p. m., giving pitch line velocities of 53 to 277 feet per minute. The third, 16 in the diagram, is a true worm, $2\frac{1}{4}$ inches pitch diameter, lead 1.333, triple thread, speed 200 to 1,442 r. p. m., bronze wheel and hardened steel worm. These figures give a pitch angle of 10° 45', and a pitch line velocity of 118 to 845 feet per minute. While this

worm is entirely successful, it was at first
a failure, and was made successful only
by careful attention to the materials used.

LIMITING SPEEDS AND PRESSURES.

A very important point connected with
worm design, and one on which data are
very scarce, is the limiting pressures for
various speeds at which cutting begins.
The paper by Mr. Lewis contains some
information on this subject, and the accompanying table supplied by Mr. Christie, from experiments made by him, sup-
plies the most definite additional data on
the subject known to the writer. In all
cases the worms were of hardened steel
and the worm wheels of cast-iron. Lu-
brication by an oil bath.

There is real need of a comprehensive
series of experiments on this subject. It
is obvious enough that a worm, otherwise
well designed, might fail from having too
high a speed for its load. Were such
data at hand it would seem that with
existing knowledge of the influence of

LIMITING SPEEDS AND PRESSURES OF WORM GEARING.

	Single Thread Worm 1″ Pitch 2⅞ Pitch Diameter			Double Thread Worm 2″ Pitch 2⅞ Pitch Diameter			Double Thread Worm 2½″ Pitch 4½ Pitch Diameter		
Revolutions per minute	128	201	272	425	128	201	272	201	272	425
Velocity at pitch line in feet per minute......	96	150	205	320	96	150	205	235	319	498
Limiting pressure in pounds	1,700	1,300	1,100	700	1,100	1,100	1,100	1,100	700	400

the angle of the thread, worm design
might be made a matter of comparative
certainty. Especially should the behavior
of worms at speeds above 200 feet per
minute be subjected to further experi-
ment, as it is frequently necessary to use
speeds above that figure, and there can
be no doubt that higher speeds are entirely feasible if suitable pressures accom-
pany them. The speed as a factor should
be kept in mind equally with the pitch
angle. A worm may fail because of too
high a pitch line velocity as well as because of too low a pitch angle.

The number of cases cited is too few
for certainty in drawing general conclu-
sions, but the testimony is unmistakable
in its confirmation of the theory of the
influence of the angle of the thread. It
will be seen that every case having an
angle above 12° 30′ was successful, and
every case below 9° unsuccessful, the
overlapping of the successful and unsuccessful worms in the intervening region
being what is to be expected in the bor-

der region between good and bad practice. This band of uncertain results is
in fact narrower than we would have any
right to expect from a collection of data
from miscellaneous sources, and could
the inquiry be widened in scope the
width of this band would doubtless be
increased. As throwing light on these
cases it should be remembered that case
16 is known to have been made successful only by careful attention to the material used, the first worms made having
been failures, and that 3, which is near
16, and was a failure, had an excessive
speed, while 11 at a lower angle, and a
success, had a very moderate speed. At
a higher speed 11 would probably have
failed, and at a lower speed 3 would probably have been a success. It is believed
that Fig. 3 points out clearly the nature
of the worm problem and the conditions
of success in its solution.

Some investigations of this subject
have been made by C. Bach and E. Roser

at the Royal Technical High School, Stuttgart, Germany, published in the "Zeitschrift des Vereines Deutscher Ingenieure," and republished in the "American Machinist" for July 16 and 23, 1903.

The worm was three-threaded, of steel, not hardened, having a total diameter of 76.6 mm. and a lead of 76.2 mm., giving a helix angle of 17° 34'. The worm wheel was of bronze with 30 teeth and a pitch diameter of 242.6 mm. The worm was immersed in an oil bath into which thermometers were inserted, and in the conduct of the experiments the apparatus was operated under various loads and speeds. Observations of the temperatures of the oil and surrounding air were made until the difference reached a constant value. Fig. 7a shows the relation of speed and pressure at two of the observed constant differences of temperature. In this diagram the temperatures are in degrees centigrade and the sliding velocities in meters per second.

The curves will be seen to be distinctly parabolic in character.

In the notation of the experimenters:

$P =$ pressure on teeth in kilograms

$t =$ pitch of teeth in centimeters.

$b =$ breadth of wheel measured on the arc at the roots of the teeth in centimeters.

$k =$ a constant.

It is then assumed that

$$P = k\,b\,t.$$

The value of k is found from the equation:

$$k = c\,(t_o - t_e) + d,$$

in which

$$c = \frac{.0669}{v} + .4192.$$

$$d = \frac{109.1}{v + 2.75} - 24.92.$$

$t_o =$ temperature of the oil in degrees centigrade.

$t_e =$ temperature of the surrounding air in degrees centigrade,

and in which again,

$v =$ sliding velocity at the pitch line in meters per second.

It is obvious, also, that the diagram furnishes a means by which the load and speed at which a worm gear works satisfactorily being known, the proper load for other speeds may be ascertained.

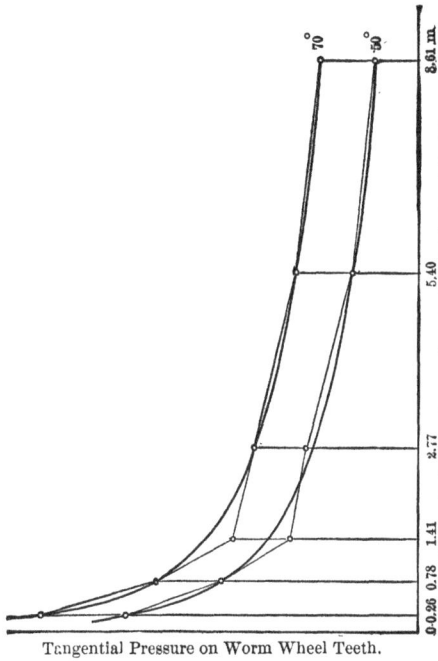

Tangential Pressure on Worm Wheel Teeth.

Sliding Velocity, Measured at the Pitch Circle of the Worm.

FIG. 7*a*. — RELATION OF PRESSURE AND VELOCITY.

STEP BEARINGS.

The step bearings of worms have been a source of trouble alongside of the worm itself. It is obvious enough that the increase of the thread angle will relieve the step of wear as well as the worm thread. Expedients are possible with the step, however, which are not available with the worm. Ball and roller step bearings have been extensively tried, and while some have been successful with them, the general results are believed to have been unsatisfactory. The troubles from ball bearings arise from the tendency of the balls to break up under heavy loads and to score the pressure plates, while conical rollers, which geometrical considerations call for, have in some instances made trouble from their outward radial pressure cutting out the confining ring.* The multiple washer thrust bear-

---

* Mr. C. R. Pratt has had marked success with roller thrust bearings in which the rollers were short cylinders kept in position by a distance plate or cage having suitable openings arranged in spirals. See " The American Machinist," June 23, 1901.

ing is used by many, and is undoubtedly
entirely successful. Many of the readers
of this volume have no doubt seen this
pattern of bearing without reflecting
upon the principle which lies at the bottom of it. When several loose washers
are interposed between the shaft collar
and the face of the shaft bearing, it is obvious that slipping may occur between
any pair of faces, and that this slipping
will take place between those surfaces
which at the moment offer the least friction. Should these surfaces from any
cause increase their resistance the slipping will be at once transferred to another joint, the various surfaces acting as
mutual safety valves to one another, any
surface which gets into the condition of
incipient heating or cutting being at
once relieved by another taking up the
work. Fig. 6 shows one of these bearings as made by the Newton Machine
Tool Works. The Newton Works formerly followed the usual practice and made
the washers *a* alternately of hardened

steel and bronze, but consider that they have improved on this by substituting white cast-iron for the steel. These castings are obtained from the malleable iron foundries, and are in fact unannealed malleable castings. Of course these castings cannot be machined, and they are therefore prepared for use by grinding on a cup-shaped emery wheel. They are dropped into a socket on the end of a shaft, which is revolved by hand, and are thus presented to the face of the cup-shaped wheel. This plan results in the grinding marks crossing the faces in all directions, instead of being in circles as in lathe-finished pieces. All mechanics understand the advantage of having the tool marks in a direction different from that of the motion of the parts, which advantage this method of construction secures. Another feature of the Newton bearings consists in making the holes in the washers larger than the shaft on which they are placed. This construction introduces an irregular compound motion of the surfaces upon one another, the advantages of which are well understood. In the Newton washers the holes are $\frac{1}{32}$-inch larger than the shaft, though Mr. Newton considers that this might be increased with probably good results.

Three radial oil grooves are cast in each face of these disks, the use of which is obvious. The Hewes & Phillips step, at the right-hand bearing, will be seen to include three washers, while the left-hand step has lenticular-shaped disks—a construction which has also found favor elsewhere.

PART II.

SPIRAL GEARING.

SPIRAL gears are not to blame for the
undoubted fact that they are somewhat
troublesome to lay out, the difficulties of
the problem being due to the limitations
of workshop facilities and not to the geometrical nature of the gears themselves.
It is easy to understand and explain the
action of an existing pair of spiral gears.
More than this it is easy to lay out a pair
of such gears which shall exactly meet all
the conditions of the case except one—
they cannot, except through rare good
luck, be made with the appliances at
hand. To be more specific, the circum-

ference cannot usually be divided into an exact whole number of teeth by any stock cutter, and the real problem becomes the readjusting of the diameters of the gears and the angle of the teeth, so that stock cutters shall make an exact whole number of teeth.

With spur gears it is only necessary to multiply the (circumferential) pitch of the cutter by the number of teeth to be cut to obtain the circumference of the gears. With spiral gears this operation gives the length of a portion of a spiral, or, more properly, helix, wound upon the pitch surface. We do not know the angle of this helix, the diameter of the pitch cylinder upon which it is wrapped or even what part of a complete turn the known portion comprises. The length is known for each gear and nothing more, and it becomes a matter of trial to find the diameters of the gears and the helix angle to suit this portion of the helix and at the same time fill the required center distance.

Fig. 8 is a conventional representation of the pitch surface of a spiral gear, the surface being extended beyond the limits of the gear in order that the two helixes with which we are concerned may be shown. The first of these, *a b c d e f*, is the tooth helix and the second, *a g h d i p*, is the normal helix. The tooth helix is of importance because it defines the angle of the teeth. Given the diameter of the pitch surface, the helix may be defined by the angle *k a l* or by the length *a f*, in which it makes a complete turn—that is, by its pitch. For the determination of the speed ratio of a pair of gears the former method is the more convenient, but the tables supplied with universal milling machines which are used in setting up the machine employ the latter method.

In all spiral gear problems we have two pitches to deal with—the pitch of the tooth helix and the pitch of the teeth. The latter may be measured in several ways. First is the value *a n*, measured on the circumference or the *circumfer-*

ential pitch, which is analogous to the pitch of spur gears; second is the value *a o* measured on the normal helix or the *normal pitch,* for which the cutters must be selected; third is the value *a r* measured parallel with the axis or the *axial pitch.** Since the cutters must be selected with reference to the normal pitch the length of the normal helix is naturally of importance in connection with the number of teeth in the gear. The normal pitch multiplied by the number of teeth must naturally equal the length *a g h d* of this helix measured between its intersections *a* and *d* with the helix of a single tooth. Note that the length of the normal helix to be considered is the length *a g h d* between its intersections with the tooth helix and not the length

---

* Of two mating gears the circumferential pitch of one is equal to the axial pitch of the other, and *vice versa.* The axial and circumferential pitches are of small importance, except in the case of worm gearing. In making a worm we deal with axial pitch, and in making a worm wheel, with circumferential pitch. In such spiral gears as are made with a formed milling cutter (not a hob) we are concerned chiefly with normal pitch.

a g h i p q of a complete turn around the
cylinder. That this is true may be seen
by reference to Fig. 9, in which the
angle *k a l* is nearly a right angle. It is
apparent from this illustration that the
length of the normal helix from *a* to *d*
takes in all the teeth and that *a o*, multiplied by the number of teeth, must equal
a h p d and not *a h p q*. This length
a h p d is always less than *a h p q*, and
usually much less. Fig. 10 *A* is a development of Fig. 9 on a reduced scale, *a d*
being the developed length of the normal
helix. Fig. 10 *B* and Fig. 10 *C* show
how with the same circumferential pitch
and the same number of teeth but a reduced value of the angle *k a l*, the length
of the normal helix which cuts all the
teeth grows shorter until it may make
but a small part of a complete turn around
the cylinder. It is clear that in all cases
the line *a d* cuts all the teeth precisely as
does the circumference *a a*, which goes
completely around the cylinder. It is
also clear that if the normal pitch is de-

cided upon at the start, a diameter of
cylinder and a helix angle must be found
such that the normal pitch, multiplied
by the number of teeth, shall equal the
length of the normal helix between two
intersections with the tooth helix.

It is natural to ask : Why not employ
the circumferential pitch and so deal directly with the circumference instead of
the normal helix ? Because we do not
know what it is. The normal pitch is
determined by the cutter used, while the
circumferential pitch depends also upon
the helix angle, and until this angle is
known the circumferential pitch is not
known.

In the extreme case of a spiral gear in
which the helix angle is so small that the
gear becomes a single thread worm, as in
Fig. 11, points o and d coincide and the
length of the helix between a and d becomes the normal pitch. It is, however,
true as before that the normal pitch,
multiplied by the number of teeth, which
is now one, is still equal to the length of

the normal helix between two intersections with the tooth helix.

A glance at Fig. 10 will show that in gears of the same diameter the length of the normal helix* grows shorter as the angle $k\,a\,l$ grows less, and hence that it and its gear will contain successively fewer and fewer teeth of the same normal pitch. That is to say, the number of teeth in a gear varies with the helix angle as well as with the diameter and *the number of teeth in two gears of the same normal pitch is not necessarily proportional to the diameters.* In fact, it is never so proportional, except when the angle $k\,a\,l$ is equal to 45 degrees. *The diametral pitch of the cutters and the diameter of the gear thus do not determine the number of teeth.*

The two facts thus developed are fundamental and will bear re-stating :

First, *The number of teeth is equal to*

---

* " Length of normal helix " is to be understood as meaning the length of that helix between two intersections with the same tooth helix.

*the length of the normal helix divided by
the normal pitch.*

Second, *The numbers of teeth in a
pair of gears are not proportional to the
diameters, except when the angle of the
tooth helix is 45 degrees.*

THE SPEED RATIO.

Fig. 12 illustrates the simplest possible
case of a pair of spiral gears. The gears
are of equal size and the tooth helix has
an angle of 45 degrees. Such a pair of
gears will obviously run at the same speed
—that is, have a speed ratio of 1—and as
obviously both will have the same number
of teeth. Now, unlike spur gears, there
are two ways in which the speed ratio of
such a pair of spiral gears may be varied.
First, the diameters of the gears may be
changed, as with spur gears, the angle of
the tooth helix remaining unchanged, as
in Fig. 13; and second, the angle of the
helix may be changed, the diameters of
the gears remaining unchanged, as in
Fig. 14. These methods act in very dif-

ferent ways. The first method is analogous to the procedure with spur gears. As with spur gears, the circumferential or pitch line speed of the two gears remains, as before the change, equal, but the length of the circumference of the two gears is unequal and the larger one thus has a less number of revolutions than the smaller one. The second method is entirely unlike anything seen in connection with spur gears. By it the pitch line speeds of the two gears are made unequal, and hence, while their diameters are equal, the lower one revolves the more slowly. This points out another fundamental difference between spiral and spur gears: With spiral gears, unless the helix angle is 45 degrees, *the pitch line speeds of two mating gears are not the same.*

The two methods of changing the speed ratio shown in Figs. 13 and 14 may be combined. That is, part of the desired change in speed may be obtained by changing the diameters of the gears and the remainder by changing the angle of the helix. Given the speed ratio and the diameter of one of the gears, we may assume a helix angle and find a diameter for the second gear to go with it which shall give the desired speed ratio and, having done this, a second angle may be assumed and a second diameter be found. There are thus an indefinite number of combinations of angles and diameters which will give the required speed ratio. Note, however, that with the diameter of one gear fixed, every change in the diameter of the other changes the distance between centers, that not every angle of helix can be obtained by the gears which are furnished with universal milling machines, and that if ready-made cutters are to be used, the lengths of both normal helixes must be exact multiples of the normal pitch of the teeth.

The limitation of the helix angle is not, however, as serious as is usually supposed. The tables for spirals which have heretofore been supplied with universal milling machines give but a few

of the spirals which can be obtained with
the change gears which are regularly
supplied with the machines. In the
case of the Brown and Sharpe universal
milling machine, about two thousand
spirals can be cut with these gears.

Geometrically speaking, there is a wide
range of choice in the helix angle. As
regards the desirability of different angles
from the standpoint of durability, the
conditions are essentially the same as in
worm gearing. Reference to Part I, will
show that the most favorable angle for
durability is at about 45 degrees. There
is, however, but a trifling increase in
wear down to 30 degrees, no serious increase down to 20 degrees, and no de-
structive increase down to about 12
degrees. Where gears are to transmit
considerable power the best results should
attend the use of angles between 30 and
45 degrees, while angles as low as 20 degrees may be used without hesitation,
and as low as 12 degrees if the gears are

to run in an oil bath or do light work
only. The angle may also be increased
above 45 degrees by similar amounts and
with similar results.

Fig. 15 is a development of the gears
of Fig. 14, the angle α of Fig. 15 being
equal to $k\,a\,l$ of Fig. 14, but in reversed
position, because in Fig. 14 the upper
side of the driver is seen, while in Fig. 15
the direction of the teeth is that of the
lower side of the driver.

It is clear that if the driver move in
the direction of the arrow it will, while
moving the distance $a\,b$, push the driven
gear the distance $b\,c$, and the pitch line
speeds will have the relation :

$$\frac{\text{p.l. speed follower}}{\text{p.l. speed driver}} = \frac{b\,c}{b\,a}$$
$$= \tan \alpha.$$

If the gears have the same diameters,
their number of revolutions will be in the
same ratio as their pitch line speeds—
that is :

$$\frac{\text{rev. follower}}{\text{rev. driver}} = \tan \alpha,$$

or

rev. follower = rev. driver $\times$ tan α

If the diameter of the follower be increased, its number of revolutions will be reduced in the same ratio—that is :

$$\text{rev. follower} = \frac{\text{diam. driver}}{\text{diam. follower}} \times$$

$$\text{rev. driver} \times \tan \alpha,$$

or

$$\frac{\text{rev. follower}}{\text{rev. driver}} = \frac{\text{diam. driver}}{\text{diam. follower}} \times \tan. \alpha,$$

the angle α being taken from the driver.

This is a complete formula for the speed ratio of spiral gears having shafts at right angles. But for the limitations imposed by the use of stock cutters it, together with the fact that the sum of the diameters of the gears must equal twice the center distance, would be all that is required for designing such gears. Note that it differs from the corresponding

formula for spur gears only by the intro-
duction of the factor tan α.

THE PRELIMINARY SOLUTION.

The simple formula for the speed given
above will be needed repeatedly, and had
best be put in algebraic form.

Let r_1 = revolutions of driver,
   r_2 = revolutions of follower,
   d_1 = diameter of driver,
   d_2 = diameter of follower,
   α = helix angle of driver.
Then this formula becomes :

$$\frac{r_2}{r_1} = \frac{d_1}{d_2} \tan \alpha \qquad (1)$$

In any actual case the speeds are given
and the diameters and helix angle must
be found. We may assume a ratio for
the diameters and find the angle, or we
may assume an angle and find the ratio
of diameters. It is desirable to assume
the angle first, as on it depends, largely,
the durability of the gears. To do this

the above formula may be more conveniently written :

$$\frac{d_2}{d_1} = \frac{r_1}{r_2} \tan \alpha \qquad (2)$$

The sum of the diameters must equal twice the center distance, which we may call C. That is :

$$d_1 + d_2 = 2C,$$

or $\qquad d_2 = 2C - d_1.$

Substituting this value for d_2 in (2) we obtain :

$$\frac{2\,C - d_1}{d_1} = \frac{r_1}{r_2} \tan \alpha$$

which, solved for d_1, becomes :

$$d_1 = \frac{2C}{\dfrac{r_1}{r_2} \tan \alpha + 1} \qquad (3)$$

Having assumed a value for α and substituted its tangent and the ratio of the desired speeds in (3), we find a value for d_1, and, having found d_1, d_2 may obviously be found by subtracting d_1 from $2\,C$.

Such a solution is complete in a geometrical sense, and if it were feasible to
make a cutter to suit each case, it would
be complete in a practical sense also.
When, however, we go a step further and
find the length of the normal helixes, the
probabilities are all against their being
exact multiples of the pitch of any stock
cutter. The solution so obtained must
therefore be considered as provisional and
be modified to suit the cutters to be used.

The table at the end of the book is
republished here from "A Treatise on
Milling Machines," by the courtesy of
the Brown and Sharpe Manufacturing
Company, and gives by inspection all
the spirals which can be obtained by the
gears supplied with the universal milling
machines provided by them. Figs. 24
and 25 show the same make of machine
with the vertical spindle milling attachment applied to the cutting of spirals
which cannot be cut in the usual manner
because of the interference of the work-
table with the column of the machine.

THE LENGTHS OF THE NORMAL HELIXES.

Fig. 16 is the development of a pair of gears placed in the most convenient position for showing the lengths of the normal helixes. The tooth and normal helixes are extended beyond the face of the gears. Let

c_1 = circumference of driver,
c_2 = circumference of follower,
d_1 = diameter of driver,
d_2 = diameter of follower,
l_1 = length of normal helix of driver between intersections with tooth helix,
l_2 = length of normal helix of follower between intersections with tooth helix,
α = tooth helix angle of driver.

Obviously

$$l_1 = c_1 \sin \alpha,$$
$$= \pi d_1 \sin \alpha,$$

or $$\frac{l_1}{\pi} = d_1 \sin \alpha \qquad (4)$$

$$l_2 = c_2 \cos \alpha,$$
$$= \pi d_2 \cos \alpha,$$

or $$\frac{l_2}{\pi} = d_2 \cos \alpha \qquad (5)$$

Note that (4) and (5) give the lengths of the normal helixes divided by π and not their actual lengths. This is done because, in dealing with diametral pitch cutters the calculations are made less laborious. Dividing (4) by (5) gives:

$$\frac{l_1}{l_2} = \frac{d_1 \sin \alpha}{d_2 \cos \alpha}$$

$$= \frac{d_1}{d_2} \tan \alpha \qquad (6)$$

Comparing (1) with (6) proves what is almost self-evident, that the lengths of the normal helixes are to each other inversely as the number of revolutions, and hence that *a pitch which will exactly divide the short helix will also divide the long one* and that *the numbers of teeth in the gears are inversely as the speeds.*

A PRACTICAL EXAMPLE.

An example will best illustrate the actual procedure. Thus assume the conditions of a pair of gears to be :

$$\frac{\text{revolutions of follower}}{\text{revolutions of driver}} = \frac{r_2}{r_1} = \tfrac{1}{4}$$

and

$$\text{center distance} = C = 4\tfrac{15}{32}$$
$$= 4.468 \text{ inches.}$$

Assume further that a helix angle of 45 degrees would lead to a diameter for the driver which would make it too small for its shaft. We are, at the start, entirely at sea regarding the whole matter; but as an angle of 30 degrees is favorable to durability we may use it as a trial angle and see what it will lead to. Finding the tangent of 30 degrees in a table and substituting it and the value of $\frac{r_1}{r_2}$ in (3) we obtain :

$$d_1 = \frac{2 \times 4.468}{4 \times .57735 + 1}$$
$$= 2.7$$

and

$$d_2 = 2 \times 4.468 - 2.7$$
$$= 6.236.$$

From (4) we find

$$\frac{l_1}{\pi} = 2.7 \times .5$$
$$= 1.35$$

and from (5)

$$\frac{l_2}{\pi} = 6.236 \times .866$$
$$= 5.4.$$

These values of d_1, d_2, $\frac{l_1}{\pi}$ and $\frac{l_2}{\pi}$ are the provisional values belonging with 30 degrees for α.

Assume next that it is desired to use cutters of 6 diametral pitch, the circumferential pitch of which is $\frac{\pi}{6} = .5236$. We may find the number of teeth which the normal helixes will contain by dividing their lengths by this circumferential pitch, but

$$l \div \frac{\pi}{6} = \frac{l}{\pi} \times 6.$$

That is, the number of teeth of 6 diametral pitch which the provisional normal helixes will contain may be found by multiplying $\frac{l_1}{\pi}$ and $\frac{l_2}{\pi}$ respectively by 6. Performing this operation we obtain :

$$\frac{l_1}{\pi} \times 6 = 1.35 \times 6$$
$$= 8.1$$

and

$$\frac{l_2}{\pi} \times 6 = 5.4 \times 6$$
$$= 32.4.$$

The provisional normal helixes thus contain 8.1 and 32.4 teeth of the desired pitch, and as these numbers are impossible, we take the nearest whole numbers having the desired ratio of 1 to 4, namely, 8 and 32. That is, we decide to shorten the normal helixes until they contain exactly 8 and 32 teeth.

The meaning of this is shown graphically in Fig. 17. Laying down ab and ac to represent the circumferences as found above and drawing the normal helixes at an angle of 30 degrees, we have found that the normal helix ad of the driver will contain 8 teeth and a little more, and the normal helix ae of the follower 32 teeth and a little more. As we cannot have a fraction of a tooth, we decide to cut off the ends of the helixes, making their lengths ad' and ae'.*

FINAL SOLUTION BY CHANGING THE CENTER DISTANCE.

The most obvious way of carrying this out is simply to reduce the diameters of both gears, so as to make their circumferences ab' and ac' instead of ab and ac. This change obviously reduces the center distance, but at this drawing-board

* Should the helixes come out a little short instead of a little long, they would be lengthened instead of shortened. Had the helix of the driver been of a length to contain, say, 8.3 teeth, that of the follower would obviously have contained 33.2 teeth, and, in the correction, more than an entire tooth would have been cut from it.

stage of affairs this can often be done,
and when it can be done it is the readiest
way out of the difficulty. To determine
how much to reduce the diameters we
may first find the reduced lengths of the
normal helixes, by dividing the final
numbers of teeth—8 and 32—by 6 and
then find the new diameters, or more
simply, we may note, what is apparent
from Fig. 17, that the lengths of the helixes and of the diameters have been
changed in the same ratio as the numbers of teeth.

That is :

$$\frac{\text{final diameter}}{\text{provisional diameter}} = \frac{8}{8.1}$$

or

final diameter = provisional diam. $\times \dfrac{8}{8.1}$

That is :

$$\text{final } d_1 = 2.7 \times \frac{8}{8.1}$$
$$= 2.667$$

and

$$\text{final } d_2 = 6.236 \times \frac{8}{8.1}$$

$$= 6.159$$

and $d_1 + d_2 = 2.667 + 6.159 = 8.825 =$ twice the new center distance.

GRAPHICAL SOLUTION WITH CHANGED CENTER DISTANCE.

All these determinations may be made graphically as in Fig. 18. Lay off $a\,b =$ $2\,C = 8\frac{15}{16}$ inches. At any convenient distance lay off the indefinite line $c\,d$ parallel to $a\,b$. At c lay off the provisional angle $\alpha = 30$ degrees. Draw $e\,f$ at any convenient point perpendicular to $c\,d$. Take $e\,f$ in the dividers and step it off from e toward d as many times as will represent the ratio of the desired speed of the driver divided by that of the follower. That is, in the present case, lay off $e\,f$ 4 times above e and thus obtain d. Draw $c\,a$ and $d\,b$ and extend them till they meet at g.* Draw $g\,e$, giving $a\,h$ and $b\,h$, which are provisional diameters of driver and follower respectively. Draw $h\,p$ perpendicular to $a\,b$, at h lay down $h\,k$ and $h\,l$ to repeat α, and from h strike arcs $a\,k$ and $b\,l$. Draw $k\,o$ and $n\,l$ perpendicular to $a\,b$, and we have the provisional values,

$$a\,h = d_1$$
$$b\,h = d_2$$
$$h\,o = \frac{l_1}{\pi}$$
$$h\,n = \frac{l_2}{\pi}$$

Scale $h\,o$ and $h\,n$ and multiply them by the diametral pitch number—6. If the results are not whole numbers, as they usually are not, select the nearest whole numbers having the desired speed ratio, and they are the final numbers of teeth. Divide these numbers by the diametral pitch number to obtain the final values

---

* Had $c\,d$ been taken shorter than $a\,b$, g would have fallen to the left of the diagram, but the constructior would otherwise have been unchanged.

of $\frac{l_1}{\pi}$ and $\frac{l_2}{\pi}$ and lay them down as $h\,o'$
and $h\,n'$. Draw $o'\,k'$ and $l'\,n'$ and $k'\,a'$
and $l'\,b'$, giving :

$a'\,h\,=\,$ the final d_1,
$b'\,h\,=\,$ the final d_2,
$a'\,b'\,=\,$ twice the new center distance.

FINAL SOLUTION WITH UNCHANGED CENTER DISTANCE.

If the center distance cannot be varied,
the condition of things is shown in Fig.
19. Having found that the provisional
normal helixes will not contain an exact
number of teeth, and cut them off and
obtained the smaller circumferences $a\,b'$
and $a\,c'$ of Fig. 17, the problem becomes
to find a new value of α, indicated in
Fig. 19, as α', which shall give such
diameters as to restore the old center distance. That is, $a\,b'' + a\,c''$ of Fig. 19
must be equal to the original $a\,b + a\,c$ of
Fig. 17. It will be seen by a glance at
Fig. 19 that the reduced value of α in-

creases the diameter of the driver, but reduces that of the follower.

It is probable that the correct angle and diameters for this case can be found only by trial. From (4) we have :

$$d_1 = \frac{l_1}{\pi \ \sin \ \alpha} \tag{7}$$

and from (5)

$$d_2 = \frac{l_2}{\pi \ \cos \ \alpha} \tag{8}$$

We also have :

$$d_1 + d_2 = 2\ C;$$

that is

$$\frac{l_1}{\pi \ \sin \ \alpha} + \frac{l_2}{\pi \ \cos \ \alpha} = 2\ C,$$

which easily reduces to :

$$1 + \frac{l_2}{l_1} \tan \ \alpha = \frac{2\ C}{\frac{l_1}{\pi}} \sin \ \alpha \tag{9}$$

To solve this equation we know that $\frac{l_2}{l_1}$ = the speed ratio = 4 and in a preceding section we found that the cor-

rected value of $\dfrac{l_1}{\pi}$ bears the same relation
to the provisional value that the corrected
number of teeth bears to the provisional
number. That is, the corrected value of
$\dfrac{l_1}{\pi} = 1.35 \times \dfrac{8}{8.1} = 1.33.$ Substituting
these values and the value of C, the only
unknown quantity is α, but, from the
form of the equation, α can be found
only by trial and error, the result of
each trial being a closer and closer ap-
proximation to the truth. The operation is simple enough, but it involves
the repeated multiplication of decimals.
For the earlier steps a slide rule will
greatly abbreviate the work, while with a
Sexton's omnimeter an accuracy can be
obtained sufficient for all the steps and a
piece of drudgery be converted into an
exhilarating chase. The results which
follow were obtained with an omnimeter.
Trying the provisional angle of 30 degrees
in order to note the result and making
the substitutions in (9) we obtain :

$$1 + 4 \times .577 = 6.702 \times .5$$

or $\qquad 3.308 = 3.351.$

This is not correct, as we knew it would not be. The right-hand number is larger than the left-hand number, which will always be the case if our trial value of α is too large, and *vice versa* if it is too small. Trying 28 degrees we obtain :

$$1 + 4 \times .532 = 6.702 \times .469$$

or $\qquad 3.13 = 3.145.$

The right-hand number is still too large, showing 28 degrees to be too large. Trying 26 degrees we obtain :

$$1 + 4 \times .488 = 6.702 \times .438$$

or $\qquad 2.953 = 2.935.$

The right-hand number is now too small, showing that 26 degrees is too small. Trying 26 degrees 30 minutes we obtain :

$$1 + 4 \times .499 = 6.702 \times .446$$

or $\qquad 3. = 2.99$

This angle is still too small, though obviously very nearly correct. Trying 26 degrees and 40 minutes we obtain :

$$1 + 4 \times .502 = 6.702 \times .449$$

or $\qquad 3.01 = 3.01$

Within the limits of accuracy of the omnimeter this is correct, and as that instrument shows the effect of so small a change in the angle as 10 minutes the result is obviously close enough. Having found the angle we are now in shape to find the final diameters. Substituting in (7) and (8) we obtain :

$$d_1 = \frac{1.333}{.4488}$$
$$= 2.971$$

and
$$d_2 = \frac{5.333}{.8936}$$
$$= 5.968$$

and $5.968 + 2.971 = 8.939$, which is twice the required center distance as closely as can be expected.

GRAPHICAL SOLUTION WITH UNCHANGED CENTER DISTANCE.

These determinations may also be made graphically, as shown in Fig. 20, which repeats the construction of Fig. 18 up to the finding of the provisional values of $a\,h$ and $b\,h$.

In Fig. 18 h was a fixed point, while a and b were not. In the present case a and b are fixed, but h is not, and to find $\frac{l_1}{\pi}$ and $\frac{l_2}{\pi}$ in the present case it is better to lay off α at a and b instead of at h, as was done in Fig. 18. Therefore lay down α at a and b, as shown, and strike arcs $h\,k$ and $h\,l$. Draw $k\,o$ and $l\,n$, giving the provisional values :

$$a\,o = \frac{l_1}{\pi}$$

$$b\,n = \frac{l_2}{\pi}$$

Measure these distances and multiply them by the diametral pitch number—in this case 6—to find the numbers of teeth going with the assumed value of α. If, as is usually the case, the products are not whole numbers, take the nearest whole numbers having the ratio of the desired speeds, divide them by the diametral pitch number and obtain the corrected values of $\dfrac{l_1}{\pi}$ and $\dfrac{l_2}{\pi}$. Lay off $a\,o'$ and $b\,n'$ equal to these corrected values. Assume a new trial angle, indicated as α', and lay it down at a and b. Draw $o'\,k'$ and $n'\cdot l'$, and through the intersections k' and l' draw the dotted arcs. If these arcs meet in tangency at h', then we have the final values :

$$a\,h' = d_1$$
and $$b\,h' = d_2$$

If the arcs do not thus meet in tangency, assume another value for α and try again.

Graphical methods are sufficiently accurate for many purposes, especially if laid out upon an enlarged scale, and if it is remembered that if the values used for

$\frac{l_1}{\pi}$ and $\frac{l_2}{\pi}$ are a little short of the true values, the shortage will be divided among the teeth with the only result of a little slackness in the teeth.

FINDING THE PITCH OF THE TOOTH HELIX.

It now only remains to find the pitch of the tooth helixes in order to set up the machine. In Fig. 16 let

p_1 = pitch of tooth helix of driver,
p_2 = pitch of tooth helix of follower;

then
$$p_1 = c_1 \tan \alpha,$$
$$= \pi d_1 \tan \alpha,$$

and
$$p_2 = c_2 \cot \alpha,$$
$$= \pi d_2 \cot \alpha.$$

Inserting the final values of d_1, d_2, and α we obtain :

$$p_1 = \pi \times 2.971 \times .502$$
$$= 4.684,$$

and
$$p_2 = \pi \times 5.968 \times 1.991$$
$$= 37.33.$$

These values may be found graphically as in Fig. 21. Lay down α and make

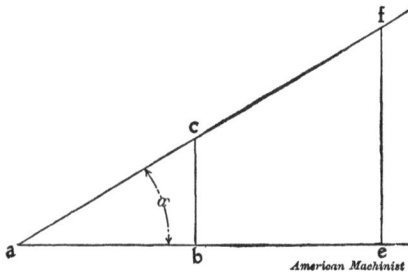

FIG. 21.—FINDING THE PITCH OF THE TOOTH HELIX.

$ab = \pi d_1$ when $bc = p_1$. Find $ef = \pi d_2$ and $ae = p_2$.

SPECIAL SOLUTION FOR A HELIX ANGLE OF 45 DEGREES.

Remembering that $\dfrac{l_2}{l_1} = \dfrac{r_1}{r_2}$, the latter fraction may be substituted for the former in (9) together with the assumed value of α and the provisional value of $\dfrac{l_1}{\pi}$

be thus found without first finding the provisional values of d_1 and d_2. The final values of d_1 and d_2 may then be found from (7) and (8). Such a solution is in some respects neater than the one given, but it is believed to be less easily followed and comprehended, and, moreover, it does not, for the general case, furnish the basis of so good a graphical solution. This method is at its best in the case of a helix angle of 45 degrees, although this angle does not differ from any other in requiring a change in the center distance if it is to be adhered to. In the nature of the case, if we deny ourselves the liberty of changing the angle, the only recourse is to correct the length of the helix by changing the diameters. As this angle is used more than any other—perhaps more than all others—this solution for it is given.

Remembering that in (9) $\dfrac{l_2}{l_1} = \dfrac{r_1}{r_2}$, that $\tan 45° = 1$, and that $\sin 45° = .7071$, that equation becomes *for this angle only:*

$$1 + \frac{r_4}{r_2} = \frac{1.4142\ C}{\dfrac{l_1}{\pi}}$$

which may be made to read :

$$\frac{l_1}{\pi} = \frac{1.4142\ C}{1 + \dfrac{r_1}{r_2}} \qquad (10)$$

Inserting sin $45° = .7071$ in (7) we have :

$$d_1 = \frac{\dfrac{l_1}{\pi}}{.7071} \qquad (11)$$

Equations (10) and (11) are all that are needed for solving any case having this helix angle.

Taking up the preceding problem again and solving it for this angle, we have from (10) for the provisional value :

$$\frac{l_1}{\pi} = \frac{1.4142 \times 4.468}{1 + 4}$$
$$= 1.2637.$$

Multiplying this by the diametral pitch of the cutter to be used, we obtain :

$$1.2637 \times 6 = 7.58$$

as the provisional number of teeth in the driver. As it is an impossible number, we select 8 as the number of teeth and, dividing it by 6, we obtain :

$$\frac{8}{6} = 1.333$$

as the final value of $\frac{l_1}{\pi}$. Substituting this in (11), we obtain :

$$d_1 = \frac{1.333}{.7071}$$
$$= 1.885$$

as the final value of the diameter of the driver. The diameter and number of teeth of the follower for an angle of 45 degrees are found by multiplying the diameter and number of teeth of the driver by the speed ratio. That is :

$$d_2 = 1.885 \times 4$$
$$= 7.540$$

and

   number of teeth in follower $= 8 \times 4$

$$= 32.$$

Finally

$d_1 + d_2 = 1.885 + 7.54$

        $= 9.425$

        $=$ twice the final center distance.

GRAPHICAL SOLUTION FOR A HELIX ANGLE OF 45 DEGREES.

The graphical solution for this angle is simplified still more than the analytical one.

In case the adjustment to suit stock cutters is to be made by changing the diameters of the gears, that adjustment can be made as in Fig. 22, which repeats essentially, Fig. 17. Laying down the provisional values of the diameters at $a\,b$ and $a\,c$ and drawing $d\,e$ at the helix angle with the base line, the perpendiculars $b\,d$ and $c\,e$ to the normal helix line determine $a\,d$ and $a\,e$, the provisional lengths of the normal helixes divided by π.

Changing these lengths to $a\,d'$ and $a\,e'$, such that when multiplied by the diametral pitch number of the cutter the results are whole numbers in the ratio of the required speeds, and drawing $d'\,b'$ and $e'\,c'$ gives $a\,b'$ and $a\,c'$, the corrected diameters.

If $\alpha = 45$ degrees the helix line becomes $f\,g$ and, with $b\,h$ perpendicular to it, $a\,b$ becomes equal to $a\,h$. In other words, instead of laying off the diameter lines at right angles to one another, they may, in this case, for purposes of determination, be laid off upon a straight line. This fact, in connection with the further fact that with a tooth angle of 45 degrees the diameters of the gears are inversely as the speeds, leads directly to the simple diagram given in Fig. 23, which is all that is required for determining graphically any pair of spiral gears having this tooth angle.

In Fig. 23 lay down $a\,b$ equal to twice the provisional center distance. Divide it at c into two parts the lengths of which

are to each other in the ratio of the
speeds. Draw de at an angle of 45 degrees with ab and draw ad and be at
right angles with de, giving cd and ce
the provisional lengths of the normal
helixes divided by π. Multiply these
lengths by the diametral pitch number.
If the results are not whole numbers take
the nearest whole numbers having the
ratio of the desired speeds, which are, respectively, the numbers of teeth in the
gears. Divide these numbers by the
diametral pitch numbers and lay down
cd' and ce' equal in inches of length to
the quotients. Draw $d'a'$ and $e'b'$ perpendicular to de, and we have :

$ca' =$ final diameter of one gear,

$cb' =$ final diameter of the other gear,

$a'b' =$ twice the final center distance.

SPECIAL SOLUTION FOR GEARS OF
EQUAL DIAMETERS.

As was pointed out on page 51, the
speed ratio of 1 which goes with gears of
equal diameters and a helix angle of
45 degrees, may be varied by changing
the diameters, the angle, or both. The
case of gears having a helix angle of
45 degrees is therefore a special and simple case in which the desired speed ratio
is obtained by recourse to a change in
diameters only. An antithetical special
and simple case is that in which the
diameters of the two gears are made
equal and the desired speed ratio is obtained by a change in the helix angle.

If this construction does not give rise
to a smaller helix angle than that which
is now well established as conducive to
durability, there is no reason why it
should not be used. This angle, it will
be remembered, is about 12 degrees,
which may be safely used for light work
and probably for heavy work if the gears

are to run in an oil bath. Under less
favorable circumstances it is not desirable to go much below 20 degrees. With
gears of equal diameters these angles
correspond to speed ratios of about $4\frac{3}{4}$
and $2\frac{3}{4}$ to 1, respectively, and it hence
follows that for these two cases and for
speed reductions not greater than these,
equal gears may be freely used.

Going back to the notation of pages 57
and 60, and letting $p =$ the diametral
pitch of the cutter to be used, formula (1),
page 57, becomes for gears of equal diameters in which $d_1 = d_2$

$$\frac{r_2}{r_1} = \tan \alpha. \qquad (12)$$

It has also been shown that

$$\frac{l_1}{\pi} = d \sin. \alpha$$

that $\qquad \dfrac{l_2}{\pi} = d \cos. \alpha$

and that the value of $\dfrac{l}{\pi}$ for either gear
multiplied by the diametral pitch equals
the number of teeth in that gear. That is:

$p\,d$ sin. α = number of teeth in driver,
and   (13)
$p\,d$ cos. α = number of teeth in follower.
                                                           (14)

If we are to use gears of equal diameters we are of course limited to tooth or helix angles appropriate to equal diamters, and as we have denied to ourselves the right to adjust the length of the normal helix by adjusting the ratio of the diameters and the helix angles, we can only do it by changing both diameters together. That is, our assumed diameters and center distance are provisional only and will ordinarily have to be changed in order to secure an exact whole number of teeth.

Practical Example: Required a pair of gears of approximately 5 inches diamter, the driver to run four times as fast as the follower; that is $\dfrac{r_2}{r_1} = \tfrac{1}{4}$.

Substituting in (12) we obtain at once tan. α = .25, and referring to a table of natural tangents, we find that $\alpha = 14° \, 2'$.

Referring to the sine and cosine columns of the table, we find that sin. $\alpha = .2425$ and cos. $\alpha = .9701$.

Assuming, say, a diametral pitch, and substituting these values in (13) and (14) we obtain for the provisional number of teeth in the driver:

$$8 \times 5 \times .2425 = 9.7$$

and in the follower:

$$8 \times 5 \times .9701 = 38.8.$$

These numbers being impossible, because fractional, we select the nearest whole numbers of teeth having the desired ratio of 4, namely, 10 and 40, and have now to find the corrected diameters to go with these tooth numbers. To do this it is only necessary to remember that the diameters are changed in the same ratio as the numbers of teeth; that is:

$$\text{Corrected diameter} = 5 \times \frac{10}{9.7}$$

or,
$$5 \times \frac{40}{38.8} = 5.155.$$

GRAPHICAL SOLUTION FOR GEARS OF
EQUAL DIAMETERS.

The graphical solution is shown in Fig. 26. Lay down ab equal to the assumed or provisional diameter. Divide it into as many parts as will represent the speed ratio—in this case 4—and draw the perpendicular bc equal to one of these parts. Draw ac and bac is the value of α sought. Extend bc and repeat α as shown. Strike the arc ad and draw de perpendicular to ab, giving $be = d\sin. \alpha$ of (13). Scale be and multiply it by the diametral pitch number 8, obtaining the provisional number of teeth in the driver, 9.7. Increase this to 10 and divide it by the diametral pitch number 8, giving 1.25, the corrected value of $d \sin. \alpha$. Lay this down from b, giving e'. Draw $e' d'$ and $d' a'$, giving the corrected diameter $a' b$. The corrected number of teeth in the follower is of course four times that in the driver, and the helix angle of the follower is the complement of that of the driver.

THE SELECTION OF THE CUTTER.[*]

There is but one more question which may be mentioned, and that is the selection of the cutter for the spiral gear. If we consider a helix drawn upon a cylinder and then imagine a plane cutting that cylinder normal to the helix at any point, the plane will cut from the cylinder an ellipse having one extremity of its minor axis at the point where the helix is cut by the plane. The minor axis of this ellipse will be $2R$ and the major axis $\frac{2R}{\cos \alpha}$. If we cut a spiral gear in the same way at a point midway between two teeth, the form of the space shown will be the true normal shape which the cutter must be, and the above ellipse will be cut from the pitch cylinder. Now the curvature of this normal section of the gear at the point indicated must be the same as the curvature of the eclipse at the extremity

---

[*] By J. N. Le Conte in the "American Machinist." The helix angle used by Professor Le Conte is the complement of the angle heretofore used in his book.

of its minor axis. If we call the major and minor axes of an ellipse $2a$ and $2b$, respectively, the radius of curvature at the extremity of the minor axis is $\rho_0 = \dfrac{a^2}{b}$.

But in our case $a = \dfrac{R}{\cos \alpha}$, and $b = R$,

Hence : $\qquad \rho_0 = \dfrac{R}{\cos^2 \alpha}$.

This is the *radius* of what might be called the "oscillating spur gear." The number of teeth on this gear will be :

$$N_0 = 2\,P\,\rho_0 = \frac{2\,P\,R}{\cos^2 \alpha}.$$

But on the spiral gear itself we have :

$$N = 2\,P\,R \cos \alpha$$

Hence : $\qquad N_0 = \dfrac{N}{\cos^3 \alpha}$

and not $\dfrac{N}{\cos^2 \alpha}$ as sometimes stated.

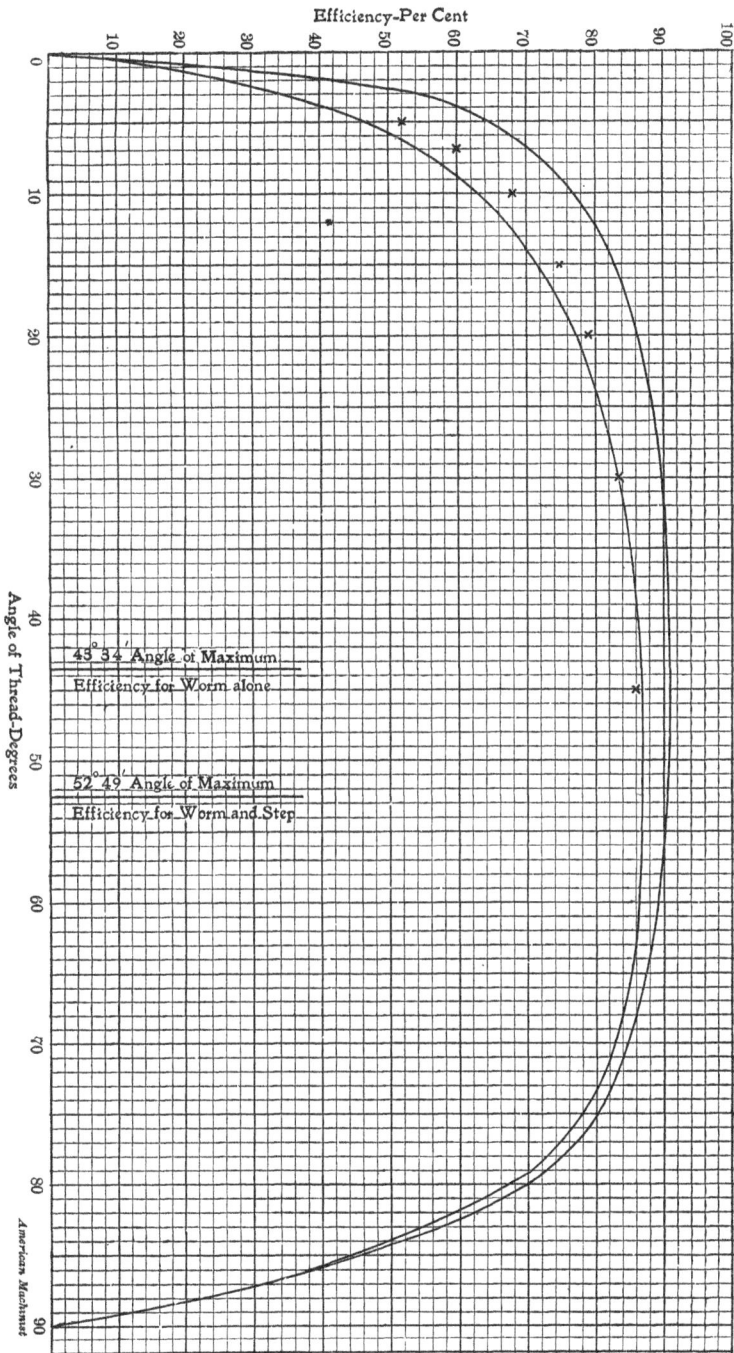

FIG. 2.—RELATION BETWEEN THREAD ANGLE AND EFFICIENCY.

Efficiency - Per Cent

Angle of Thread-Degrees

Velocity at Pitch Line
in Feet per Minute

American Machinist

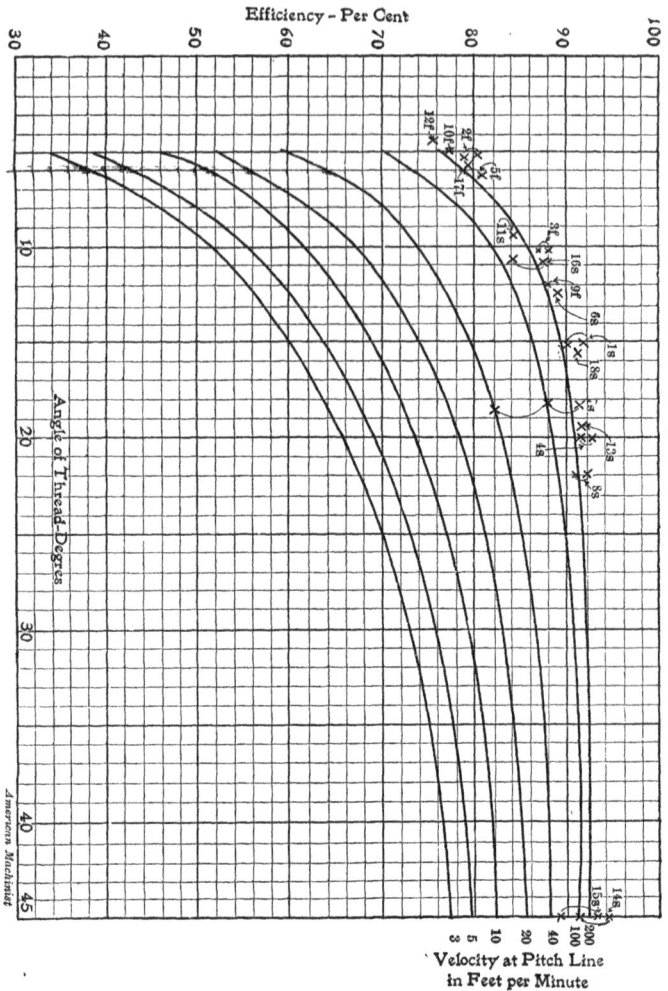

FIG. 3.—RELATION BETWEEN THREAD ANGLE, SPEED, AND EFFICIENCY,
WITH CASES FROM PRACTICE.

1. Hewes & Phillips, successful, velocity 237 and 500 feet, steel worm and cast-iron gear.
2. Hewes & Phillips, failure, velocity.... 730 " 1,780 " " " " "
3. Hewes & Phillips, failure, velocity.... 452 " 1,130 " " " " "
4. Hewes & Phillips, successful, velocity 205 " 480 " " " " "
5. Newton Mach. Tool Works, failure, velocity............ 572 " " " " bronze
6. Newton Mach. Tool Works, successful, velocity....... 375 " " " " "
7. Newton Mach. Tool Works, successful, velocity........ 40 to 685 " " " " "
8. Bertram & Sons, successful, velocity........ 155 and 620 " cast-iron " bronze
9. Anonymous, failure, velocity ... 250 " steel " cast-iron
10. Christie, failure, velocity... 215 " " " cast-iron
11. Christie, successful, velocity 190 " " " "
12. Christie, failure, velocity.... 775 " " " bronze
13. Christie, successful, velocity 328 " " " "
14. Anonymous, successful, velocity 116 to 555 " cast-iron " cast iron
15. Anonymous, successful, velocity 53 " 277 " " " "
16. Anonymous, velocity 118 " 860 " steel " bronze
17. Hunt, failure, velocity.... 296 " material unknown
18. Hunt, successful, velocity 271 " " "

Leaders connecting crosses indicate the same worm at different speeds;
distance of crosses above 200 foot line has no quantitative significance.

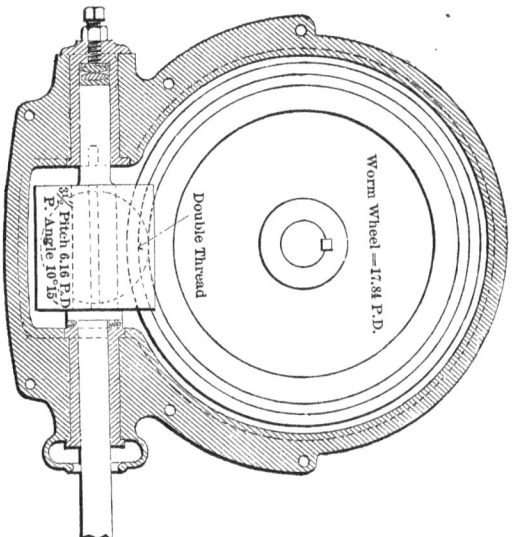

FIG. 4.
HEWES & PHILLIPS' UNSUCCESSFUL WORM.

Worm Wheel = 17.84 P.D.

Double Thread

3¾ Pitch 6.16 P.D.
P. Angle 10°15′

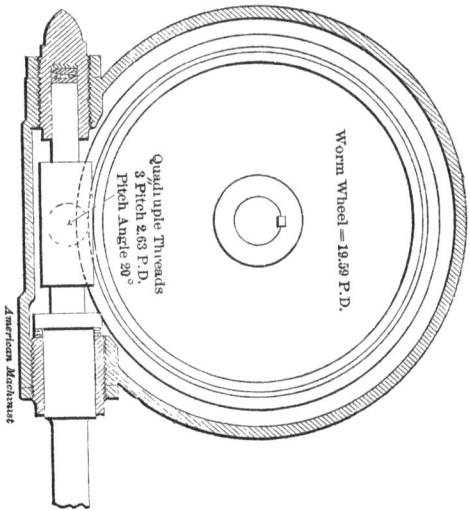

FIG. 5.
HEWES & PHILLIPS' SUCCESSFUL WORM.

Worm Wheel = 19.59 P.D.

Quadruple Threads
3 Pitch 2.63 P.D.
Pitch Angle 30°

American Machinist

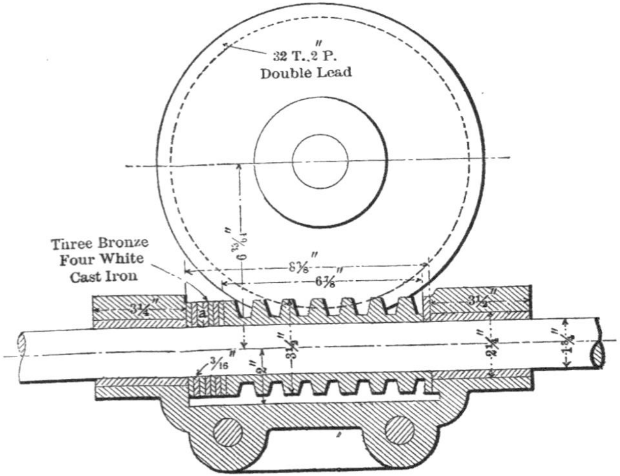
32 T., 2" P.
Double Lead
Three Bronze
Four White
Cast Iron
6 13/64"
8⅜"
6⅞"
3¼"
3½"
3/16"
2"
3⅛"
2¼"
1⅜"
FIG. 6.—THE NEWTON WORM AND STEP.

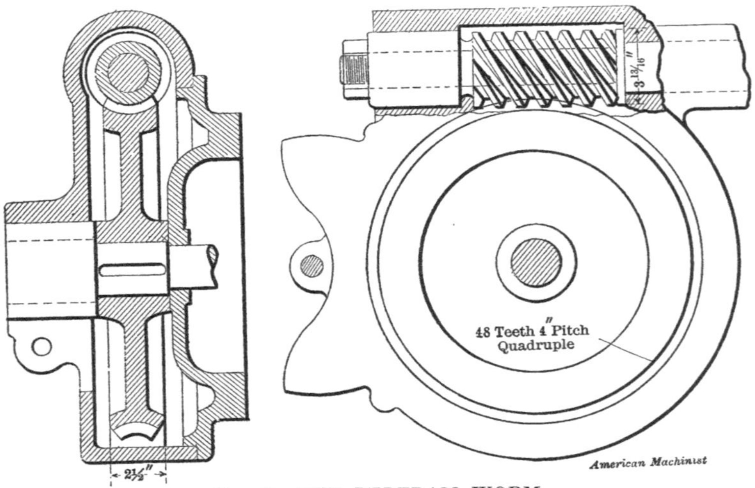
2½"
3 13/16"
48 Teeth 4" Pitch
Quadruple
American Machinist
FIG. 7.—THE BERTRAM WORM.

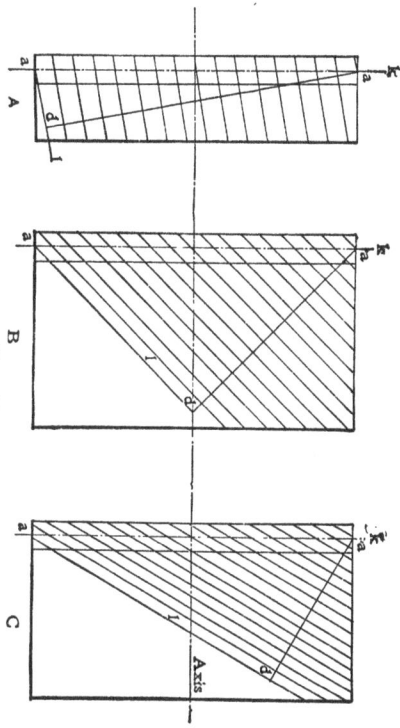

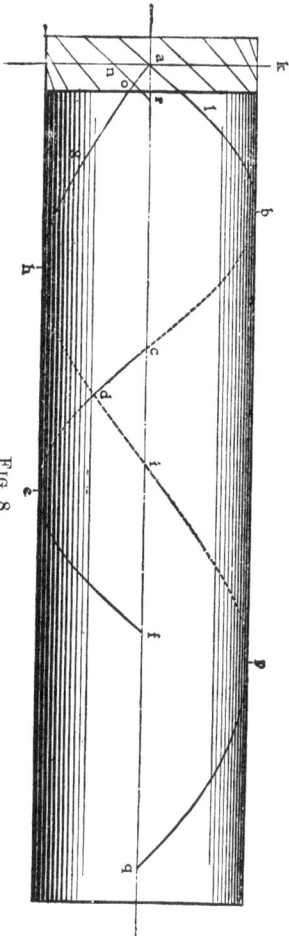

FIG. 8.

FIG. 10.

TOOTH AND NORMAL HELICES WITH THEIR DEVELOPMENT.

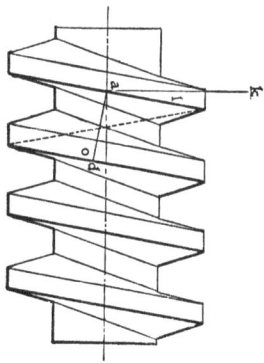

FIG. 11.

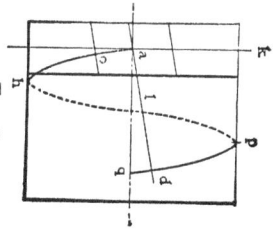

FIG. 9.

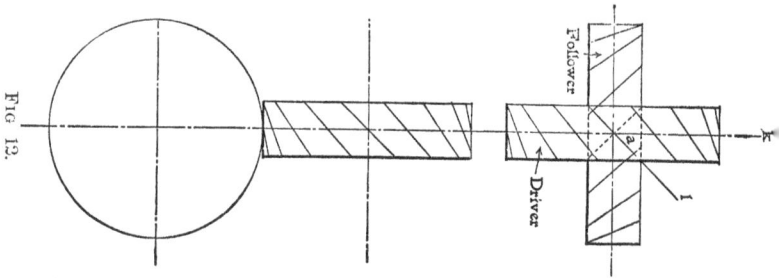

FIG. 12.

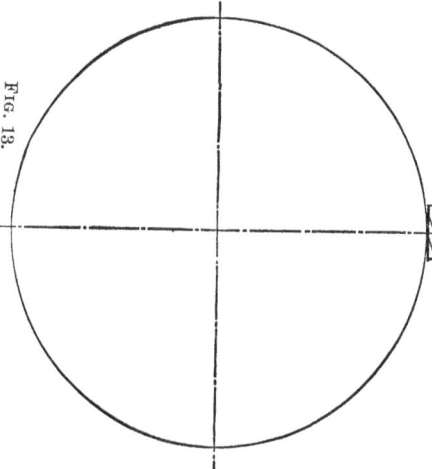

FIG. 13.

THE SPEED RATIO.

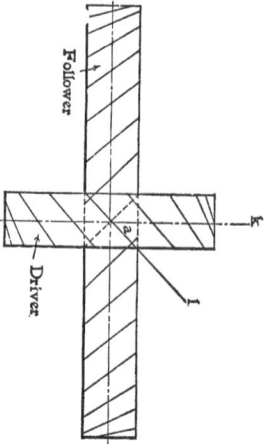

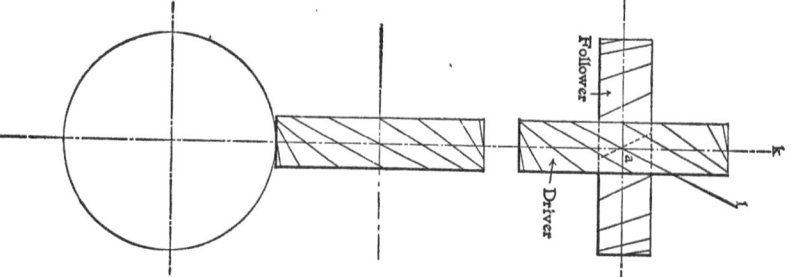

FIG. 14.

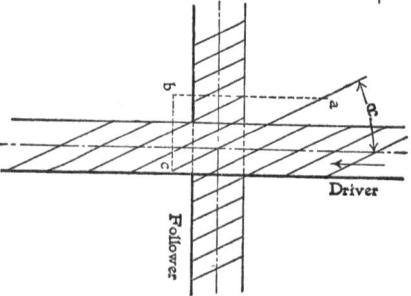

FIG. 15.

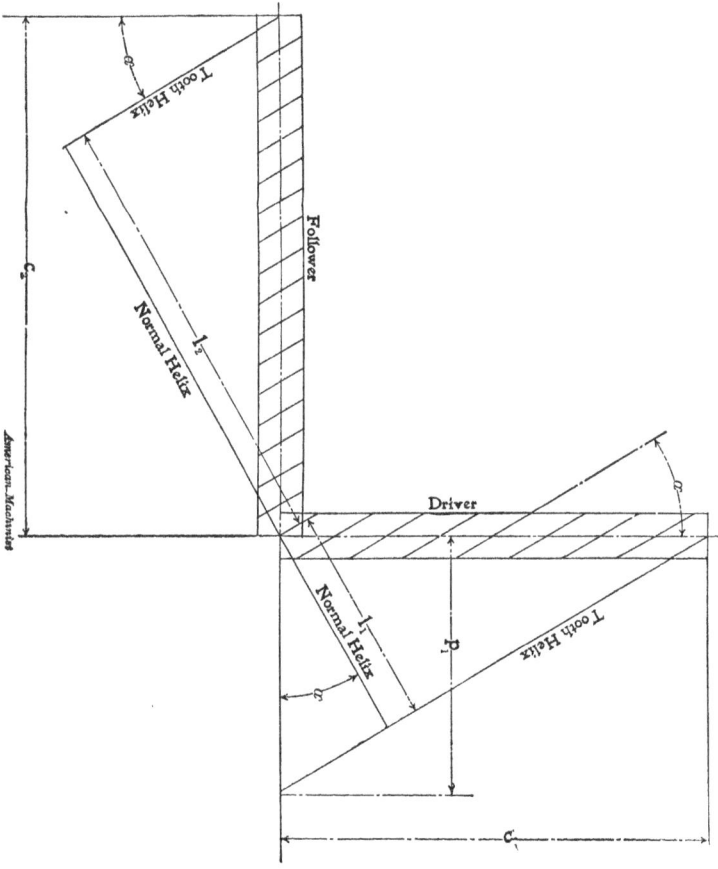

FIG. 16.—A PAIR OF GEARS DEVELOPED.

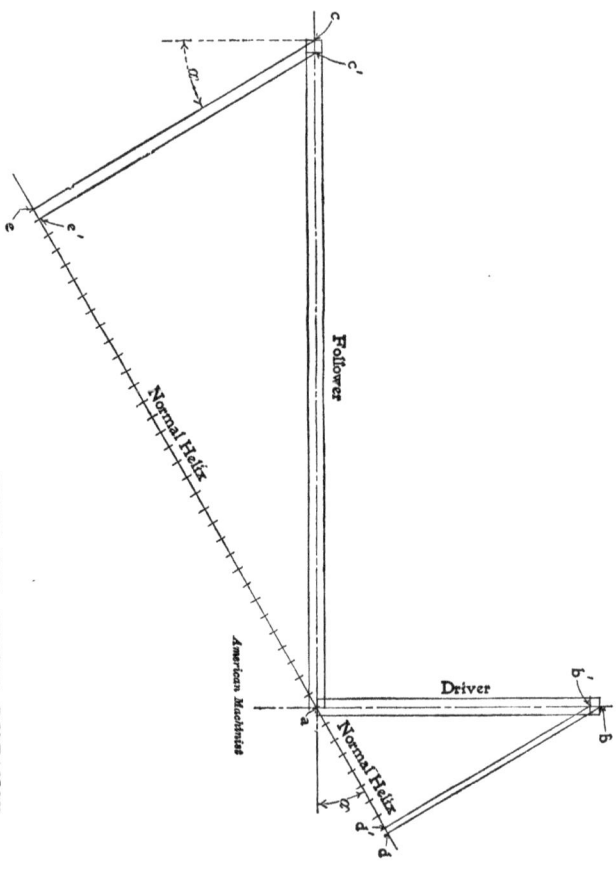

FIG. 17.—ADJUSTING THE DIAMETERS WHEN THE CENTER DISTANCE CAN BE CHANGED.

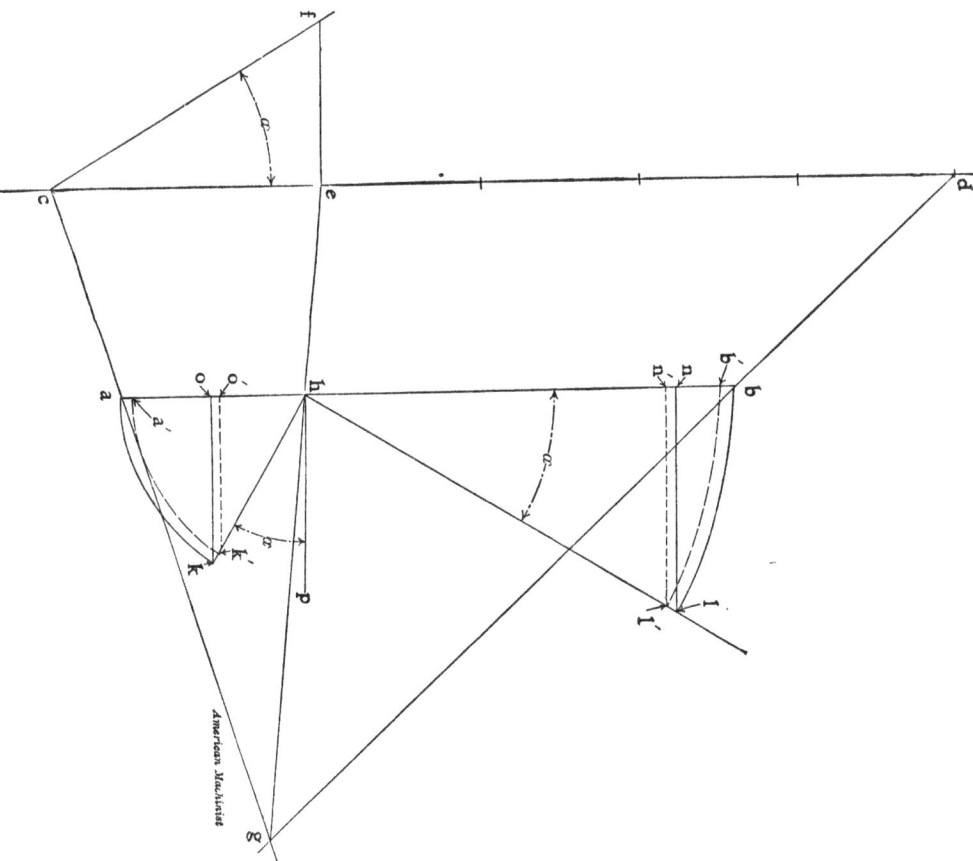

FIG. 18. GRAPHICAL SOLUTION WITH VARIABLE CENTER DISTANCE.

American Machinist

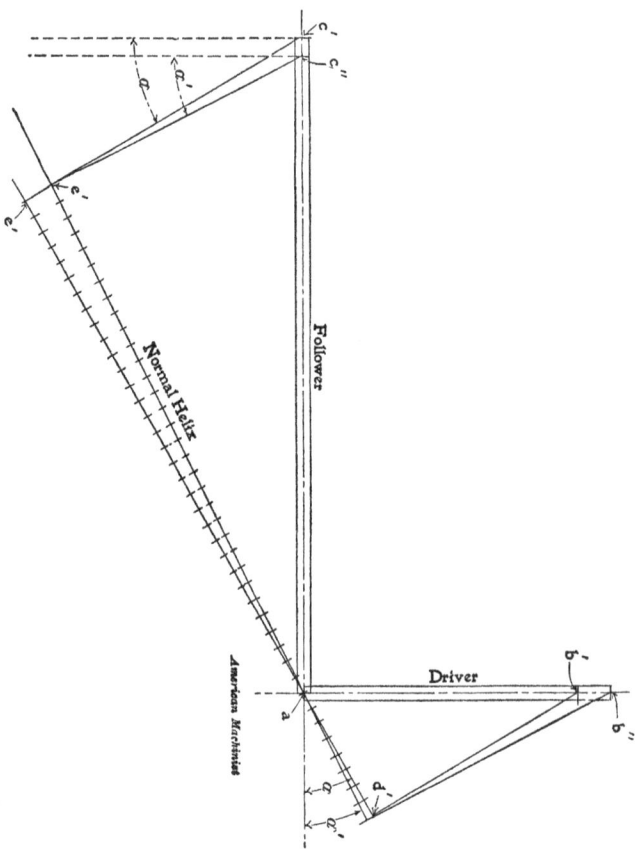

Fig. 19.—Adjusting the diameters when the center distance is fixed.

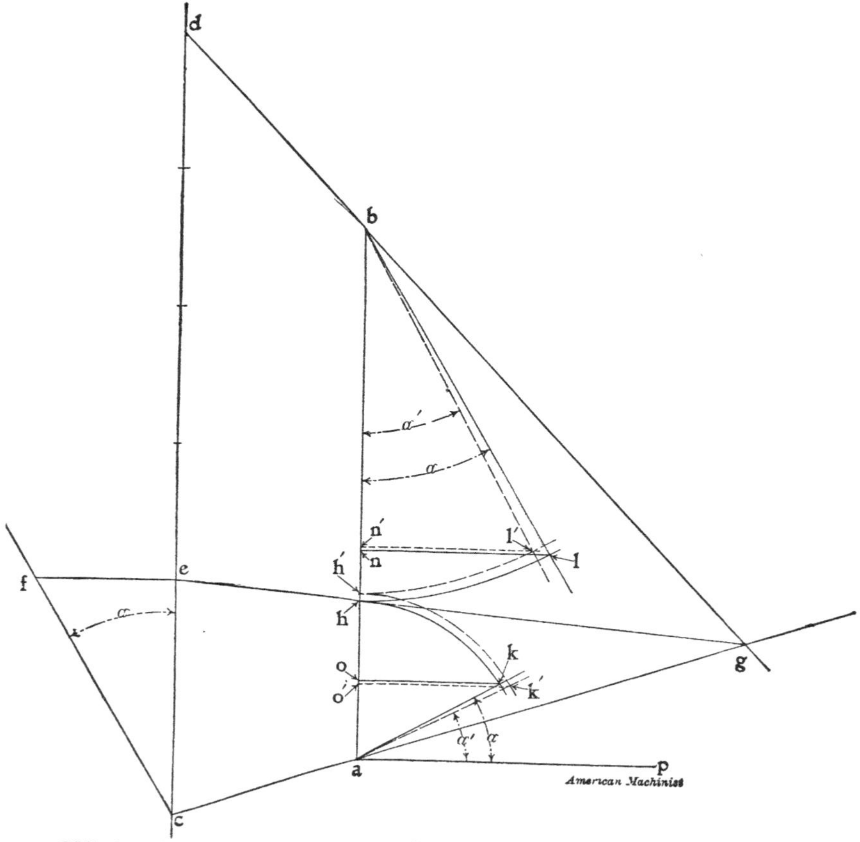

FIG. 20.—GRAPHICAL SOLUTION WITH FIXED CENTER DISTANCE.

American Machinist

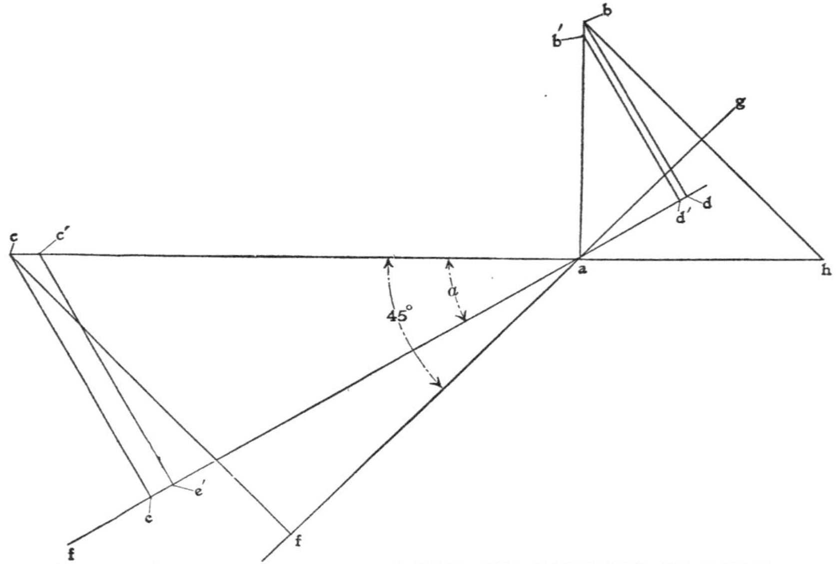

FIG. 22.—RELATION OF NORMAL HELIX AND DIAMETER IN SPIRAL
GEARS HAVING A TOOTH ANGLE OF 45 DEGREES.

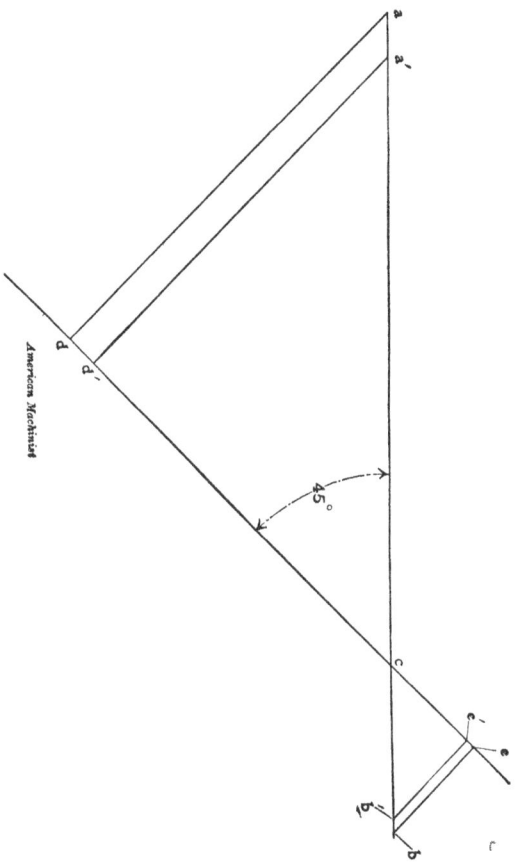

American Machinist

FIG. 23.—SIMPLE GRAPHICAL SOLUTION FOR SPIRAL GEARS HAVING A
TOOTH ANGLE OF 45 DEGREES.

45°

FIG. 24.

FIG. 25.

CUTTING SPIRALS IN THE BROWN AND SHARPE MILLING MACHINE WITH THE AID OF THE VERTICAL SPINDLE MILLING ATTACHMENT.

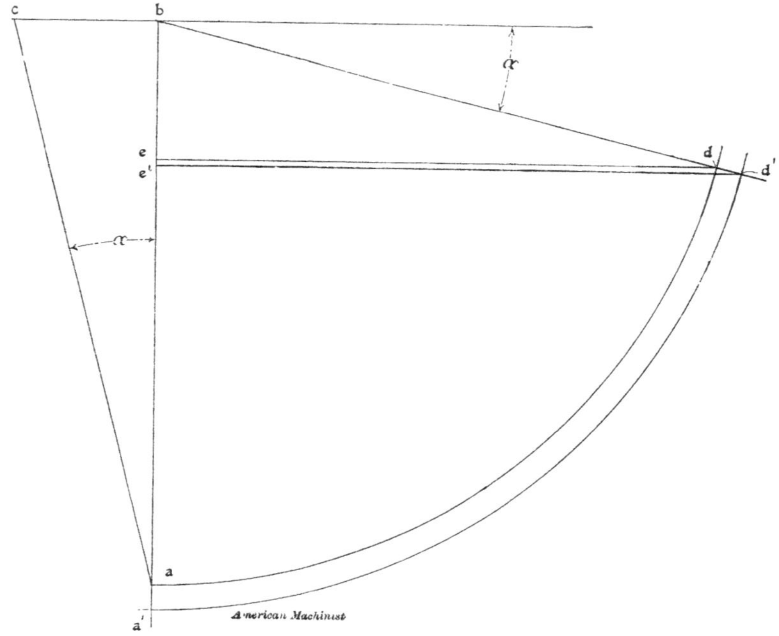

FIG. 26.—SIMPLE GRAPHICAL SOLUTION FOR SPIRAL GEARS
OF EQUAL DIAMETERS.

TABLE OF LEADS.

The following table gives all leads that can be obtained with any possible combinations of the change gears furnished with the Brown & Sharpe Universal Milling Machines.

In compiling this table all the combinations are given even though some are not available for use on the machines. Where one combination of gears will not reach or the gears interfere, another can usually be found, giving the same, or approximately the same, lead.

Combinations of gears that are too small in diameter to reach for right-hand spirals can generally be used for left-hand spirals, as the reverse gear is required and will enable the gears to reach.

The gears of any combination can be transposed; that is, the two driving gears, the two driven gears, or both, but not a driver substituted for a driven, or *vice versa.* This is important and should be thoroughly understood in order that the table may be used to the best advantage.

Example : Required, the gears to give a lead of 3.60″.

Referring to the table, the gears called for are as follows : Gear on screw, first driver, 100 teeth. 1st gear on stud, second driver, 32 teeth. 2nd gear on stud, first driven, 24 teeth. Gear on worm, second driven, 48 teeth.

By transposing the gears the following four combinations can be obtained :

DRIVING GEARS.

	1st	2d	3d	4th
Gear on Screw	100	32	100	32
1st Gear on Stud	32	100	32	100

DRIVEN GEARS.

	1st	2d	3d	4th
2d Gear on Stud	24	24	48	48
Gear on Worm	48	48	24	24

The first combination, however, is the only one available, owing to the interference of the gears in the others preventing their meshing properly.

Leads that are not given in the table can many times be approximated.

For example : Required, the gears to to give a lead of 10.855″.

Referring to the table, it will be found that this lead cannot be obtained exactly with the regular gears furnished with the machine. The nearest approximations are found to be 10.853″, which is .002″ short, and 10.859″, being .004″ too long. Either one of these, however, would be nearer than ordinary accuracy would require.

	DRIVEN	DRIVER	DRIVEN	DRIVER		DRIVEN	DRIVER	DRIVEN	DRIVER		DRIVEN	DRIVER	DRIVEN	DRIVER
LEAD IN INCHES	GEAR ON WORM	1ST GEAR ON STUD	2ND GEAR ON STUD	GEAR ON SCREW	LEAD IN INCHES	GEAR ON WORM	1ST GEAR ON STUD	2ND GEAR ON STUD	GEAR ON SCREW	LEAD IN INCHES	GEAR ON WORM	1ST GEAR ON STUD	2ND GEAR ON STUD	GEAR ON SCREW
.670	24	86	24	100	1.527	24	44	28	100	1.886	24	56	44	100
.781	24	86	28	100	1.550	24	72	40	86	1.905	24	56	32	72
.800	24	72	24	100	1.556	28	72	40	100	1.919	24	64	44	86
.893	24	86	32	100	1.563	24	86	56	100	1.920	24	40	32	100
.900	24	64	24	100	1.563	28	86	48	100	1.925	28	64	44	100
.930	24	72	24	86	1.595	24	56	32	86	1.944	24	48	28	72
.933	24	72	28	100	1.600	24	48	32	100	1.944	28	64	32	72
1.029	24	56	24	100	1.600	28	56	32	100	1.954	24	40	28	86
1.042	28	86	32	100	1.600	24	72	48	100	1.956	32	72	44	100
1.047	24	64	24	86	1.607	24	56	24	64	1.990	28	72	44	86
1.050	24	64	28	100	1.628	24	48	28	86	1.993	24	56	40	86
1.067	24	72	32	100	1.628	28	64	32	86	2.000	24	40	24	72
1.085	24	72	28	86	1.637	32	86	44	100	2.000	24	48	40	100
1.116	24	86	40	100	1.650	24	64	44	100	2.000	28	56	40	100
1.196	24	56	24	86	1.667	24	56	28	72	2.000	32	64	40	100
1.200	24	48	24	100	1.667	24	48	24	72	2.009	24	86	72	100
1.200	24	56	28	100	1.667	24	64	32	72	2.030	24	44	32	86
1.200	24	64	32	100	1.674	24	40	24	86	2.035	28	64	40	86
1.221	24	64	28	86	1.680	24	40	28	100	2.036	28	44	32	100
1.228	24	86	44	100	1.706	24	72	44	86	2.045	24	44	24	64
1.240	24	72	32	86	1.711	28	72	44	100	2.047	40	86	44	100
1.244	28	72	32	100	1.714	24	56	40	100	2.057	24	28	24	100
1.250	24	64	24	72	1.744	24	64	40	86	2.057	24	56	48	100
1.302	28	86	40	100	1.745	24	44	32	100	2.067	32	72	40	86
1.309	24	44	24.	100	1.750	28	64	40	100	2.083	24	64	40	72
1.333	24	72	40	100	1.776	24	44	28	86	2.084	28	86	64	100
1.340	24	86	48	100	1.778	32	72	40	100	2.084	32	86	56	100
1.371	24	56	32	100	1.786	24	86	64	100	2.093	24	64	48	86
1.395	24	48	24	86	1.786	32	86	48	100	2.093	24	32	24	86
1.395	24	56	28	86	1.800	24	64	48	100	2.100	24	64	56	100
1.395	24	64	32	86	1.800	24	32	24	100	2.100	28	64	48	100
1.400	24	48	28	100	1.809	28	72	40	86	2.100	24	32	28	100
1.400	28	64	32	100	1.818	24	44	24	72	2.121	24	44	28	72
1.429	24	56	24	72	1.823	28	86	56	100	2.133	24	72	64	100
1.433	28	86	44	100	1.860	28	56	32	86	2.133	32	72	48	100
1.440	24	40	24	100	1.861	24	72	48	86	2.143	24	56	32	64
1.447	28	72	32	86	1.861	24	48	32	86	2.143	24	48	24	56
1.458	24	64	28	72	1.867	28	48	32	100	2.171	24	72	56	86
1.467	24	72	44	100	1.867	24	72	56	100	2.171	28	48	32	86
1.488	32	86	40	100	1.867	28	72	48	100	2.171	28	72	48	86
1.500	24	64	40	100	1.875	24	48	24	64	2.178	28	72	56	100
1.522	24	44	24	86	1.875	24	56	28	64	2.182	24	44	40	100

DRIVEN GEAR ON WORM	DRIVER 1ST GEAR ON STUD	DRIVEN 2ND GEAR ON STUD	DRIVER GEAR ON SCREW	LEAD IN INCHES	DRIVEN GEAR ON WORM	DRIVER 1ST GEAR ON STUD	DRIVEN 2ND GEAR ON STUD	DRIVER GEAR ON SCREW	LEAD IN INCHES	DRIVEN GEAR ON WORM	DRIVER 1ST GEAR ON STUD	DRIVEN 2ND GEAR ON STUD	DRIVER GEAR ON SCREW
2.188	24	48	28	64	2.500	24	48	28	56	2.800	24	24	28	100
2.193	24	56	44	86	2.500	28	56	32	64	2.800	32	64	56	100
2.200	24	48	44	100	2.500	24	64	48	72	2.800	24	48	56	100
2.200	28	56	44	100	2.500	24	48	32	64	2.812	24	32	24	64
2.200	32	64	44	100	2.500	24	32	24	72	2.828	28	44	32	72
2.222	24	48	32	72	2.514	32	56	44	100	2.843	40	72	44	86
2.222	28	56	32	72	2.532	28	72	56	86	2.845	32	72	64	100
2.233	40	86	48	100	2.537	24	44	40	86	2.849	28	64	56	86
2.233	24	40	32	86	2.546	28	44	40	100	2.857	24	48	32	56
2.238	28	64	44	86	2.558	32	64	44	86	2.857	24	56	48	72
2.240	28	40	32	100	2.558	28	56	44	86	2.857	24	28	24	72
2.250	24	40	24	64	2.558	24	48	44	86	2.865	44	86	56	100
2.274	32	72	44	86	2.567	28	48	44	100	2.867	86	72	24	100
2.286	32	56	40	100	2.571	24	40	24	56	2.880	24	40	48	100
2.292	24	64	44	72	2.593	28	48	32	72	2.894	28	72	64	86
2.326	32	64	40	86	2.605	28	40	32	86	2.894	32	72	56	86
2.326	24	48	40	86	2.605	40	86	56	100	2.909	32	44	40	100
2.326	28	56	40	86	2.618	24	44	48	100	2.917	24	64	56	72
2.333	28	48	40	100	2.619	24	56	44	72	2.917	28	64	48	72
2.333	24	40	28	72	2.625	24	40	28	64	2.917	28	48	32	64
2.338	24	44	24	56	2.640	24	40	44	100	2.917	24	32	28	72
2.344	28	86	72	100	2.658	32	56	40	86	2.924	32	56	44	86
2.368	28	44	32	86	2.667	40	72	48	100	2.933	44	72	48	100
2.381	32	86	64	100	2.667	32	48	40	100	2.934	32	48	44	100
2.381	24	56	40	72	2.667	24	40	32	72	2.946	24	56	44	64
2.386	24	44	28	64	2.674	28	64	44	72	2.950	28	44	40	86
2.392	24	56	48	86	2.678	24	56	40	64	2.977	40	86	64	100
2.392	24	28	24	86	2.679	32	86	72	100	2.984	28	48	44	86
2.400	28	56	48	100	2.700	24	64	72	100	3.000	24	40	28	56
2.400	32	64	48	100	2.713	28	48	40	86	3.000	24	40	32	64
2.424	24	44	32	72	2.727	24	44	32	64	3.000	24	32	40	100
2.431	28	64	40	72	2.727	24	44	28	56	3.000	40	64	48	100
2.442	24	32	28	86	2.727	24	44	24	48	3.000	24	40	24	48
2.442	28	64	48	86	2.743	24	56	64	100	3.030	24	44	40	72
2.442	24	64	56	86	2.743	32	56	48	100	3.044	24	44	48	86
2.445	40	72	44	100	2.743	24	28	32	100	3.055	28	44	48	100
2.450	28	64	56	100	2.750	40	64	44	100	3.055	24	44	56	100
2.456	44	86	48	100	2.778	32	64	40	72	3.056	32	64	44	72
2.481	32	72	48	86	2.778	24	48	40	72	3.056	28	56	44	72
2.481	24	72	64	86	2.778	40	56	28	72	3.056	24	48	44	72
2.489	32	72	56	100	2.791	28	56	48	86	3.070	24	40	44	86
2.489	28	72	64	100	2.791	32	64	48	86	3.080	28	40	44	100

LEAD IN INCHES	GEAR ON WORM	1ST GEAR ON STUD	2ND GEAR ON STUD	GEAR ON SCREW	LEAD IN INCHES	GEAR ON WORM	1ST GEAR ON STUD	2ND GEAR ON STUD	GEAR ON SCREW	LEAD IN INCHES	GEAR ON WORM	1ST GEAR ON STUD	2ND GEAR ON STUD	GEAR ON SCREW
	DRIVEN	DRIVER	DRIVEN	DRIVER		DRIVEN	DRIVER	DRIVEN	DRIVER		DRIVEN	DRIVER	DRIVEN	DRIVER
3.086	24	56	72	100	3.349	48	40	24	86	3.637	48	44	24	72
3.101	40	72	48	86	3.360	56	40	24	100	3.646	40	48	28	64
3.101	32	48	40	86	3.360	48	40	28	100	3.655	40	56	44	86
3.111	28	40	32	72	3.383	32	44	40	86	3.657	64	56	32	100
3.111	40	72	56	100	3.403	28	64	56	72	3.663	72	64	28	86
3.117	24	44	32	56	3.409	24	44	40	64	3.667	40	48	44	100
3.125	28	56	40	64	3.411	32	48	44	86	3.667	44	40	24	72
3.125	24	48	40	64	3.411	44	72	48	86	3.673	24	28	24	56
3.126	48	86	56	100	3.422	44	72	56	100	3.684	44	86	72	100
3.140	24	86	72	64	3.428	24	40	32	56	3.686	86	56	24	100
3.143	40	56	44	100	3.429	40	28	24	100	3.704	32	48	40	72
3.150	28	100	72	64	3.429	40	56	48	100	3.721	24	24	32	86
3.175	32	56	40	72	3.438	24	48	44	64	3.721	64	48	24	86
3.182	28	44	32	64	3.438	28	56	44	64	3.721	64	56	28	86
3.182	24	44	28	48	3.488	40	64	48	86	3.733	48	72	56	100
3.189	32	56	48	86	3.488	40	32	24	86	3.733	56	48	32	100
3.189	24	28	32	86	3.491	64	44	24	100	3.733	64	48	28	100
3.190	24	86	64	56	3.491	48	44	32	100	3.733	28	24	32	100
3.198	40	64	44	86	3.492	32	56	44	72	3.750	24	32	24	48
3.200	28	100	64	56	3.500	40	64	56	100	3.750	24	32	28	56
3.200	24	100	64	48	3.500	28	32	40	100	3.750	28	56	48	64
3.200	24	24	32	100	3.500	28	40	32	64	3.763	86	64	28	100
3.214	24	56	48	64	3.500	24	40	28	48	3.771	44	56	48	100
3.214	24	32	24	56	3.520	32	40	44	100	3.772	24	28	44	100
3.214	24	28	24	64	3.535	28	44	40	72	3.799	56	48	28	86
3.225	24	100	86	64	3.552	56	44	24	86	3.809	24	28	32	72
3.241	28	48	40	72	3.552	48	44	28	86	3.810	64	56	24	72
3.256	24	24	28	86	3.556	40	72	64	100	3.810	32	56	48	72
3.256	24	86	56	48	3.564	56	44	28	100	3.818	24	40	28	44
3.256	32	64	56	86	3.565	28	48	44	72	3.819	40	64	44	72
3.267	28	48	56	100	3.571	24	48	40	56	3.822	86	72	32	100
3.273	24	40	24	44	3.571	32	56	40	64	3.837	24	32	44	86
3.275	44	86	64	100	3.572	48	86	64	100	3.837	44	64	48	86
3.281	24	32	28	64	3.582	44	40	28	86	3.840	64	40	24	100
3.300	44	64	48	100	3.588	72	56	24	86	3.840	32	40	48	100
3.300	44	32	24	100	3.600	72	48	24	100	3.850	44	64	56	100
3.308	32	72	64	86	3.600	72	64	32	100	3.850	28	32	44	100
3.333	32	64	48	72	3.600	72	56	28	100	3.876	24	72	100	86
3.333	28	56	48	72	3.600	48	32	24	100	3.889	32	64	56	72
3.333	28	48	32	56	3.618	56	72	40	86	3.889	56	48	24	72
3.345	28	100	86	72	3.636	24	44	32	48	3.889	24	24	28	72
3.349	40	86	72	100	3.636	28	44	32	56	3.896	24	44	40	56

LEAD IN INCHES	DRIVEN GEAR ON WORM	DRIVER 1ST GEAR ON STUD	DRIVEN 2ND GEAR ON STUD	DRIVER GEAR ON SCREW	LEAD IN INCHES	DRIVEN GEAR ON WORM	DRIVER 1ST GEAR ON STUD	DRIVEN 2ND GEAR ON STUD	DRIVER GEAR ON SCREW	LEAD IN INCHES	DRIVEN GEAR ON WORM	DRIVER 1ST GEAR ON STUD	DRIVEN 2ND GEAR ON STUD	DRIVER GEAR ON SCREW
3.907	28	40	48	86	4.200	48	64	56	100	4.480	56	40	32	100
3.907	56	40	24	86	4.200	56	32	24	100	4.480	64	40	28	100
3.911	44	72	64	100	4.200	28	32	48	100	4.500	72	64	40	100
3 920	28	40	56	100	4.200	72	48	28	100	4.500	48	40	24	64
3.927	72	44	24	100	4.242	28	44	32	48	4.500	24	32	24	40
3.929	32	56	44	64	4.242	28	44	48	72	4.522	100	72	28	86
3.929	24	48	44	56						4.537	56	48	28	72
3.977	28	44	40	64	4.242	24	44	56	72	4.545	24	44	40	48
3.979	44	72	56	86	4.253	64	56	32	86	4.546	28	44	40	56
3.987	24	28	40	86	4.264	40	48	44	86	4.546	32	44	40	64
3.987	40	56	48	86	4.267	64	48	32	100	4.548	44	72	64	86
4.000	24	40	32	48	4.267	48	72	64	100	4.558	56	40	28	86
4.000	28	40	32	56	4.278	28	40	44	72	4.567	72	44	24	86
4.000	24	24	40	100	4.286	24	28	24	48	4.572	40	56	64	100
4.000	24	40	48	72	4.286	24	28	32	64	4.572	32	28	40	100
4.011	28	48	44	64	4.286	32	56	48	64	4.582	72	44	28	100
4.019	72	100	48	86	4.300	86	56	28	100	4.583	44	64	48	72
4.040	32	44	40	72	4.300	86	64	32	100	4.583	44	32	24	72
4.059	32	44	48	86	4.300	86	48	24	100	4.584	32	48	44	64
4.066	64	44	24	86	4.320	72	40	24	100	4.584	28	48	44	56
4.070	28	32	40	86	4.341	48	72	56	86	4.651	40	24	24	86
4.070	40	64	56	86	4.341	56	48	32	86	4.655	64	44	32	100
4.073	64	44	28	100	4.342	64	48	28	86	4.667	28	40	32	48
4.073	56	44	32	100	4.342	28	24	32	86	4.667	40	24	28	100
4.074	32	48	44	72	4.361	100	64	24	86	4.667	56	40	24	72
4.091	24	44	48	64	4.363	24	40	32	44	4.667	48	40	28	72
4.091	24	32	24	44	4.364	40	44	48	100	4.667	40	48	56	100
4.093	32	40	44	86	4.365	40	56	44	72	4.675	24	28	24	44
4.114	48	28	24	100	4.375	24	24	28	64	4.675	48	44	24	56
4.114	72	56	32	100	4.375	24	32	28	48	4.687	40	32	24	64
4.125	24	40	44	64	4.375	56	48	24	64	4.688	56	86	72	100
4 135	40	72	64	86	4.386	24	28	44	86	4.691	86	44	24	100
4.144	56	44	28	86	4.386	44	56	48	86	4.714	44	40	24	56
4.167	28	48	40	56	4.400	24	24	44	100	4.736	64	44	28	86
4.167	40	64	48	72	4.444	64	56	28	72	4.736	56	44	32	86
4.167	32	48	40	64	4.444	24	24	32	72	4.762	40	28	24	72
4.167	24	32	40	72	4.444	64	48	24	72	4.762	40	48	32	56
4.167	56	86	64	100	4.465	64	40	24	86	4.762	40	56	48	72
4.186	72	64	32	86	4.466	48	40	32	86	4.773	24	32	28	44
4.186	48	32	24	86	4.477	44	32	28	86	4.773	56	44	24	64
4.186	72	48	24	86	4.477	56	64	44	86	4 773	48	44	28	64
4.186	72	56	28	86	4.479	86	64	24	72	4.778	86	72	40	100

LEAD IN INCHES	DRIVEN GEAR ON WORM	DRIVER 1ST GEAR ON STUD	DRIVEN 2ND GEAR ON STUD	DRIVER GEAR ON SCREW	LEAD IN INCHES	DRIVEN GEAR ON WORM	DRIVER 1ST GEAR ON STUD	DRIVEN 2ND GEAR ON STUD	DRIVER GEAR ON SCREW	LEAD IN INCHES	DRIVEN GEAR ON WORM	DRIVER 1ST GEAR ON STUD	DRIVEN 2ND GEAR ON STUD	DRIVER GEAR ON SCREW
4.784	72	56	32	86	5.116	44	24	24	86	5.358	64	86	72	100
4.785	48	28	24	86	5.119	86	56	24	72	5.375	86	64	40	100
4.800	48	24	24	100	5.120	64	40	32	100	5.400	72	32	24	100
4.800	56	28	24	100	5.133	56	48	44	100	5.400	72	64	48	100
4.800	64	32	24	100	5.134	44	24	28	100	5.413	64	44	32	86
4.800	72	48	32	100	5.142	72	56	40	100	5.426	40	24	28	86
4.813	44	40	28	64	5.143	24	28	24	40	5.427	40	48	56	86
4.821	72	56	24	64	5.143	24	40	48	56	5.444	56	40	28	72
4.849	32	44	48	72	5.156	44	32	24	64	5.455	48	44	28	56
4.849	64	44	24	72	5.160	86	40	24	100	5.455	32	44	48	64
4.861	40	32	28	72	5.168	100	72	32	86	5.469	40	32	28	64
4.861	56	64	40	72	5.185	28	24	32	72	5.473	86	44	28	100
4.884	48	64	56	86	5.186	64	48	28	72	5.486	64	28	24	100
4.884	72	48	28	86	5.186	56	48	32	72	5.486	48	28	32	100
4.884	48	32	28	86	5.195	32	44	40	56	5.486	48	56	64	100
4.884	56	32	24	86	5.209	100	64	24	72	5.500	44	40	24	48
4.889	32	40	44	72	5.210	64	40	28	86	5.500	44	40	32	64
4.898	24	28	32	56	5.210	56	40	32	86	5.500	40	32	44	100
4.900	56	32	28	100	5.226	86	64	28	72	5.500	44	40	28	56
4.911	40	56	44	64	5.233	72	64	40	86	5.556	40	24	24	72
4.914	86	56	32	100	5.236	72	44	32	100	5.568	56	44	28	64
4.950	56	44	28	72	5.238	44	28	24	72	5.581	64	32	24	86
4.950	72	64	44	100	5.238	32	48	44	56	5.581	56	28	24	86
4.961	64	48	32	86	5.238	44	56	48	72	5.581	72	48	32	86
4.961	64	72	48	86	5.250	24	32	28	40	5.582	48	24	24	86
4.978	56	72	64	100	5.250	56	40	24	64	5.600	56	24	24	100
4.984	100	56	24	86	5.250	48	40	28	64	5.600	48	24	28	100
5.000	24	24	28	56	5.256	86	72	44	100	5.600	64	32	28	100
5.000	24	24	32	64	5.280	48	40	44	100	5.625	48	32	24	64
5.000	48	32	24	72	5.303	28	44	40	48	5.625	72	48	24	64
5.017	86	48	28	100	5.316	40	28	32	86	5.625	72	56	28	64
5.023	72	40	24	86	5.316	40	56	64	86	5.657	56	44	32	72
5.029	44	28	32	100	5.328	72	44	28	86	5.657	72	56	44	100
5.029	64	56	44	100	5.333	40	24	32	100	5.657	64	44	28	72
5.040	72	40	28	100	5.333	64	40	24	72	5.698	56	32	28	86
5.074	40	44	48	86	5.333	32	40	48	72	5.714	48	28	24	72
5.080	64	56	32	72	5.333	40	48	64	100	5.714	24	28	32	48
5.088	100	64	28	86	5.347	44	64	56	72	5.714	24	24	32	56
5.091	56	44	40	100	5.348	44	32	28	72	5.714	64	48	24	56
5.091	28	40	32	44	5.357	40	28	24	64	5.730	40	48	44	64
5.093	40	48	44	72	5.357	40	32	24	56	5.733	86	48	32	100
5.105	28	48	56	64	5.357	40	56	48	64	5.733	86	72	48	100

DRIVEN	DRIVER	DRIVEN	DRIVER		DRIVEN	DRIVER	DRIVEN	DRIVER		DRIVEN	DRIVER	DRIVEN	DRIVER
LEAD IN INCHES	GEAR ON WORM	1ST GEAR ON STUD	2ND GEAR ON STUD	GEAR ON SCREW	LEAD IN INCHES	GEAR ON WORM	1ST GEAR ON STUD	2ND GEAR ON STUD	GEAR ON SCREW	LEAD IN INCHES	GEAR ON WORM	1ST GEAR ON STUD	2ND GEAR ON STUD	GEAR ON SCREW
5.756	72	64	44	86	6.089	72	44	32	86	6.417	44	40	28	48
5.759	86	56	24	64	6.109	56	44	48	100	6.429	24	28	24	32
5.760	72	40	32	100	6.112	24	24	44	72	6.429	48	28	24	64
5.788	64	72	56	86	6.122	40	28	24	56	6.429	48	32	24	56
5.814	100	64	32	86	6.125	56	40	28	64	6.429	72	48	24	56
5.814	100	56	28	86	6.137	72	44	24	64	6.429	72	56	32	64
5.814	100	48	24	86	6.140	48	40	44	86	6.450	86	64	48	100
5.818	64	44	40	100	6.143	86	56	40	100	6.450	86	32	24	100
5.833	28	24	24	48	6.160	56	40	44	100	6.460	100	72	40	86
5.833	32	24	28	64				...		6.465	64	44	32	72
5.833	56	32	24	72	6.171	72	56	48	100	6.482	56	48	40	72
5.833	48	32	28	72	6.172	72	28	24	100	6.482	40	24	28	72
5.833	56	48	32	64	6.202	40	24	32	86	6.512	56	24	24	86
5.833	56	64	48	72	6.202	64	48	40	86	6.512	64	32	28	86
5.847	64	56	44	86	6.222	64	40	28	72	6.512	48	24	28	86
5.848	44	28	32	86	6.222	56	40	32	72	6.515	86	44	24	72
5.861	72	40	28	86	6.234	32	28	24	44	6.534	56	24	28	100
5.867	44	24	32	100	6.234	64	44	24	56	6.545	48	40	24	44
5.867	64	48	44	100	6.234	48	44	32	56	6.545	72	44	40	100
5.893	44	32	24	56	6.250	24	24	40	64	6.548	44	48	40	56
5.893	44	28	24	64	6.250	40	32	24	48	6.563	56	32	24	64
5.893	48	56	44	64	6.250	40	32	28	56	6.563	72	48	28	64
5.912	86	64	44	100	6.255	86	44	32	100	6.563	48	32	28	64
5.920	56	44	40	86	6.279	72	64	48	86	6.578	72	56	44	86
5.926	64	48	32	72	6.279	72	32	24	86	6.600	48	32	44	100
5.952	100	56	24	72	6.286	44	40	32	56	6.600	72	48	44	100
5.954	64	40	32	86	6.286	44	28	40	100	6.645	100	56	32	86
5.969	44	24	28	86	6.300	72	32	28	100	6.667	64	48	28	56
5.969	56	48	44	86	6.300	72	64	56	100	6.667	32	24	28	56
5.972	86	48	24	72	6.343	100	44	24	86	6.667	32	24	24	48
5.972	86	56	28	72	6.350	40	28	32	72	6.667	48	24	24	72
5.972	86	64	32	72	6.350	64	56	40	72	6.667	56	28	24	72
5.980	72	56	40	86	6.364	56	44	24	48	6.667	64	32	24	72
6.000	48	40	28	56	6.364	56	44	32	64	6.689	86	72	56	100
6.000	48	40	32	64	6.364	24	24	28	44	6.697	100	56	24	64
6.000	48	32	40	100	6.379	64	28	24	86	6.698	72	40	32	86
6.000	72	48	40	100	6.379	48	28	32	86	6.719	86	48	24	64
6.016	44	32	28	64	6.379	64	56	48	86	6.719	86	56	28	64
6.020	86	40	28	100	6.396	44	32	40	86	6.720	56	40	48	100
6.061	40	44	32	48	6.400	64	24	24	100	6.735	44	28	24	56
6.061	48	44	40	72	6.400	48	24	32	100	6.750	72	40	24	64
6.077	100	64	28	72	6.400	56	28	32	100	6.757	86	56	44	100

LEAD IN INCHES	GEAR ON WORM (DRIVEN)	1ST GEAR ON STUD (DRIVER)	2ND GEAR ON STUD (DRIVEN)	GEAR ON SCREW (DRIVER)
6.766	64	44	40	86
6.784	100	48	28	86
6.806	56	32	28	72
6.818	40	32	24	44
6.818	48	44	40	64
6.822	44	24	32	86
6.822	64	48	44	86
6.825	86	56	32	72
6.857	32	28	24	40
6.857	64	40	24	56
6.857	48	40	32	56
6.857	48	28	40	100
6.875	44	24	24	64
6.875	44	32	24	48
6.875	44	32	28	56
6.880	86	40	32	100
6.944	100	48	24	72
6.944	100	64	32	72
6.945	100	56	28	72
6.968	86	48	28	72
6.977	48	32	40	86
6.977	100	40	24	86
6.977	72	48	40	86
6.982	64	44	48	100
6.984	44	28	32	72
6.984	64	56	44	72
7.000	28	24	24	40
7.000	56	40	24	48
7.000	56	40	32	64
7.000	56	32	40	100
7.013	72	44	24	56
7.040	64	40	44	100
7.071	56	44	40	72
7.104	56	44	48	86
7.106	100	72	44	86
7.111	64	40	32	72
7.130	44	24	28	72
7.130	56	48	44	72
7.143	40	28	32	64
7.143	40	28	24	48
7.143	40	24	24	56

LEAD IN INCHES	GEAR ON WORM (DRIVEN)	1ST GEAR ON STUD (DRIVER)	2ND GEAR ON STUD (DRIVEN)	GEAR ON SCREW (DRIVER)
7.159	72	44	28	64
7.163	56	40	44	86
7.167	86	40	24	72
7.167	86	48	40	100
7.176	72	28	24	86
7.176	72	56	48	86
7.200	72	24	24	100
7.268	100	64	40	86
7.272	64	44	28	56
7.273	32	24	24	44
7.273	64	44	24	48
7.292	56	48	40	64
7.292	40	32	28	48
7.292	40	24	28	64
7.310	44	28	40	86
7.314	64	28	32	100
7.326	72	32	28	86
7.326	72	64	56	86
7.330	86	44	24	64
7.333	44	24	40	100
7.333	48	40	44	72
7.334	44	40	32	48
7.347	48	28	24	56
7.371	86	56	48	100
7.372	86	28	24	100
7.400	100	44	28	86
7.408	40	24	32	72
7.408	64	48	40	72
7.424	56	44	28	48
7.442	64	24	24	86
7.442	48	24	32	86
7.442	56	28	32	86
7.465	86	64	40	72
7.467	64	24	28	100
7.467	56	24	32	100
7.467	64	48	56	100
7.500	48	24	24	64
7.500	56	28	24	64
7.500	48	32	28	56
7.500	72	48	28	56
7.500	72	48	32	64

LEAD IN INCHES	GEAR ON WORM (DRIVEN)	1ST GEAR ON STUD (DRIVER)	2ND GEAR ON STUD (DRIVEN)	GEAR ON SCREW (DRIVER)
7.525	86	32	28	100
7.525	86	64	56	100
7.543	48	28	44	100
7.576	100	44	24	72
7.597	56	24	28	86
7.601	86	44	28	72
7.611	72	44	40	86
7.619	64	48	32	56
7.619	64	56	48	72
7.620	64	28	24	72
7.620	48	28	32	72
7.636	56	40	24	44
7.636	48	40	28	44
7.639	44	32	40	72
7.644	86	72	64	100
7.657	56	32	28	64
7.674	72	48	44	86
7.675	48	32	44	86
7.679	86	48	24	56
7.679	86	56	32	64
7.680	64	40	48	100
7.700	56	32	44	100
7.714	72	40	24	56
7.752	100	48	32	86
7.752	100	72	48	86
7.778	32	24	28	48
7.778	56	24	24	72
7.778	48	24	28	72
7.778	64	32	28	72
7.792	40	28	24	44
7.792	48	44	40	56
7.813	100	48	24	64
7.813	100	56	28	64
7.815	56	40	48	86
7.818	86	44	40	100
7.838	86	48	28	64
7.855	72	44	48	100
7.857	44	24	24	56
7.857	44	28	24	48
7.872	44	28	32	64
7.875	72	40	28	64
7.883	86	48	44	100

	DRIVEN	DRIVER	DRIVEN	DRIVER		DRIVEN	DRIVER	DRIVEN	DRIVER		DRIVEN	DRIVER	DRIVEN	DRIVER
LEAD IN INCHES	GEAR ON WORM	1ST GEAR ON STUD	2ND GEAR ON STUD	GEAR ON SCREW	LEAD IN INCHES	GEAR ON WORM	1ST GEAR ON STUD	2ND GEAR ON STUD	GEAR ON SCREW	LEAD IN INCHES	GEAR ON WORM	1ST GEAR ON STUD	2ND GEAR ON STUD	GEAR ON SCREW
7.920	72	40	44	100	8.333	48	32	40	72	8.772	48	28	44	86
7.936	100	56	32	72	8.333	100	40	24	72	8.800	48	24	44	100
7.954	40	32	28	44	8.334	40	24	28	56	8.800	64	32	44	100
7.955	56	44	40	64	8.361	86	40	28	72	8.800	56	28	44	100
7.963	86	48	32	72	8.372	72	24	24	86	8.838	100	44	28	72
7.974	48	28	40	86	8.377	86	44	24	56	8.839	72	56	44	64
7.994	100	64	44	86	8.400	72	24	28	100	8.889	64	24	24	72
8.000	64	32	40	100	8.400	56	32	48	100	8.889	56	28	32	72
8.000	32	24	24	40	8.400	72	48	56	100	8.889	48	24	32	72
8.000	64	40	24	48	8.437	72	32	24	64	8.909	56	40	28	44
8.000	64	40	28	56	8.457	100	44	32	86	8.929	100	48	24	56
8.000	56	28	40	100	8.484	32	24	28	44	8.929	100	56	32	64
8.000	48	24	40	100	8.485	64	44	28	48	8.930	64	40	48	86
8.021	44	32	28	48	8.485	56	44	32	48	8.953	56	32	44	86
8.021	44	24	28	64	8.485	56	44	48	72	8.959	86	48	28	56
8.021	56	48	44	64	8.506	64	28	32	86	8.959	86	32	24	72
8.035	72	56	40	64	8.523	100	44	24	64	8.959	86	64	48	72
8.063	86	40	24	64	8.527	44	24	40	86	8.959	86	48	32	64
8.081	64	44	40	72	8.532	86	56	40	72	8.960	64	40	56	100
8.102	100	48	28	72	8.534	64	24	32	100	8.980	44	28	32	56
8.119	64	44	48	86	8.552	86	44	28	64	9.000	48	32	24	40
8.140	56	32	40	86	8.556	56	40	44	72	9.000	72	40	24	48
8.140	100	40	28	86	8.572	64	32	24	56	9.000	72	40	28	56
8.145	64	44	56	100	8.572	48	28	32	64	9.000	72	40	32	64
8.148	64	48	44	72	8.572	48	24	24	56	9.000	72	32	40	100
8.149	44	24	32	72	8.572	72	48	32	56	9.044	100	72	56	86
8.163	40	28	32	56	8.594	44	32	40	64	9.074	56	24	28	72
8.167	56	40	28	48	8.600	86	24	24	100	9.091	40	24	24	44
8.182	48	32	24	44	8.640	72	40	48	100	9.115	100	48	28	64
8.182	72	44	24	48	8.681	100	64	40	72	9.134	72	44	48	86
8.182	72	44	28	56	8.682	64	24	28	86	9.137	100	56	44	86
8.182	72	44	32	64	8.682	56	24	32	86	9.143	64	40	32	56
8.186	64	40	44	86	8.682	64	48	56	86	9.143	64	28	40	100
8.212	86	64	44	72	8.687	86	44	32	72	9.164	72	44	56	100
8.229	72	28	32	100	8.721	100	32	24	86	9.167	44	24	24	48
8.229	72	56	64	100	8.721	100	64	48	86	9.167	44	24	28	56
8.250	44	32	24	40	8.727	48	40	32	44	9.167	44	24	32	64
8.250	48	40	44	64	8.730	44	28	40	72	9.167	48	32	44	72
8.306	100	56	40	86	8.750	28	24	24	32	9.210	72	40	44	86
8.312	64	44	32	56	8.750	56	32	24	48	9.214	86	40	24	56
8.333	40	24	24	48	8.750	56	24	24	64	9.260	100	48	32	72
8.333	40	24	32	64	8.750	48	24	28	64	9.302	48	24	40	86

LEAD IN INCHES	DRIVEN GEAR ON WORM	DRIVER 1ST GEAR ON STUD	DRIVEN 2ND GEAR ON STUD	DRIVER GEAR ON SCREW	LEAD IN INCHES	DRIVEN GEAR ON WORM	DRIVER 1ST GEAR ON STUD	DRIVEN 2ND GEAR ON STUD	DRIVER GEAR ON SCREW	LEAD IN INCHES	DRIVEN GEAR ON WORM	DRIVER 1ST GEAR ON STUD	DRIVEN 2ND GEAR ON STUD	DRIVER GEAR ON SCREW
9.303	56	28	40	86	9.675	86	64	72	100	10.101	100	44	32	72
9.303	64	32	40	86	9.690	100	48	40	86	10.159	64	28	32	72
9.303	100	40	32	86	9.697	64	48	32	44	10.175	100	32	28	86
9.333	64	40	28	48	9.697	64	44	48	72	10.175	100	64	56	86
9.333	56	40	32	48	9.723	40	24	28	48	10.182	64	40	28	44
9.333	56	24	40	100	9.723	56	32	40	72	10.182	56	40	32	44
9.333	56	40	48	72	9.723	100	40	28	72	10.186	44	24	40	72
9.334	32	24	28	40	9.741	100	44	24	56	10.209	56	24	28	64
9.351	48	28	24	44	9.768	72	48	56	86	10.209	56	32	28	48
9.351	72	44	32	56	9.768	56	32	48	86	10.228	72	44	40	64
9.375	48	32	40	64	9.768	72	24	28	86	10.233	48	24	44	86
9.375	100	40	24	64	9.773	86	44	24	48	10.233	56	28	44	86
9.375	72	48	40	64	9.773	86	44	28	56	10.233	64	32	44	86
9.382	86	44	48	100	9.773	86	44	32	64	10.238	86	28	24	72
9.385	86	56	44	72	9.778	64	40	44	72	10.238	86	48	32	56
9.406	86	40	28	64	9.796	64	28	24	56	10.238	86	56	48	72
9.428	44	28	24	40	9.796	48	28	32	56	10.267	56	24	44	100
9.429	48	40	44	56	9.818	72	40	24	44	10.286	48	28	24	40
9.460	86	40	44	100	9.822	44	32	40	56	10.286	72	40	32	56
9.472	64	44	56	86	9.822	44	28	40	64	10.286	72	28	40	100
9.524	40	28	32	48	9.828	86	28	32	100	10.312	48	32	44	64
9.524	40	24	32	56	9.828	86	56	64	100	10.313	72	48	44	64
9.524	48	28	40	72	9.844	72	32	28	64	10.320	86	40	48	100
9.524	64	48	40	56	9.900	72	32	44	100	10.336	100	72	64	86
9.545	72	44	28	48	9.921	100	56	40	72	10.370	64	24	28	72
9.546	56	32	24	44	9.923	64	24	32	86	10.370	56	24	32	72
9.546	48	32	28	44	9.943	100	44	28	64	10.371	64	48	56	72
9.547	56	44	48	64	9.954	86	48	40	72	10.390	40	28	32	44
9.549	100	64	44	72	9.967	100	56	48	86	10.390	64	44	40	56
9.556	86	40	32	72	9.968	100	28	24	86	10.417	100	32	24	72
9.569	72	28	32	86	10.000	56	28	24	48	10.417	100	48	28	56
9.569	72	56	64	86	10.000	48	24	28	56	10.417	100	48	32	64
9.598	86	56	40	64	10.000	64	32	24	48	10.417	100	64	48	72
9.600	72	24	32	100	10.000	64	32	28	56	10.419	64	40	56	86
9.600	56	28	48	100	10.000	56	28	32	64	10.451	86	32	28	72
9.600	64	32	48	100	10.000	48	24	32	64	10.451	86	64	56	72
9.600	72	48	64	100	10.033	86	24	28	100	10.467	72	32	40	86
9.625	44	32	28	40	10.033	86	48	56	100	10.473	72	44	64	100
9.625	56	40	44	64	10.046	72	40	48	86	10.476	44	24	32	56
9.643	72	32	24	56	10.057	64	28	44	100	10.476	44	28	32	48
9.643	72	28	24	64	10.078	86	32	24	64	10.477	48	28	44	72
9.643	72	56	48	64	10.080	72	40	56	100	10.477	64	48	44	56

LEAD IN INCHES	GEAR ON WORM (DRIVEN)	1ST GEAR ON STUD (DRIVER)	2ND GEAR ON STUD (DRIVEN)	GEAR ON SCREW (DRIVER)	LEAD IN INCHES	GEAR ON WORM (DRIVEN)	1ST GEAR ON STUD (DRIVER)	2ND GEAR ON STUD (DRIVEN)	GEAR ON SCREW (DRIVER)	LEAD IN INCHES	GEAR ON WORM (DRIVEN)	1ST GEAR ON STUD (DRIVER)	2ND GEAR ON STUD (DRIVEN)	GEAR ON SCREW (DRIVER)
10.500	56	32	24	40	11.111	48	24	40	72	11.667	64	32	28	48
10.500	48	32	28	40	11.111	56	28	40	72	11.667	56	32	48	72
10.500	72	40	28	48	11.111	64	32	40	72	11.667	56	24	32	64
10.500	56	40	48	64	11.111	100	40	32	72	11.688	72	44	40	56
10.558	86	56	44	64	11.137	56	32	28	44	11.695	64	28	44	86
10.571	100	44	40	86	11.160	100	56	40	64	11.719	100	32	24	64
10.606	56	44	40	48	11.163	72	24	32	86	11.721	72	40	56	86
10.606	40	24	28	44	11.163	56	28	48	86	11.728	86	40	24	44
10.631	64	28	40	86	11.163	72	48	64	86	11.733	64	24	44	100
10.655	72	44	56	86	11.163	64	32	48	86	11.757	86	32	28	64
10.659	100	48	44	86	11.169	86	44	32	56	11.785	72	48	44	56
10.667	64	40	48	72	11.198	86	48	40	64	11.786	44	28	24	32
10.667	64	24	40	100	11.200	56	24	48	100	11.786	48	32	44	56
10.667	64	40	32	48	11.200	64	32	56	100	11.786	48	28	44	64
10.694	44	24	28	48	11.225	44	28	40	56	11.825	86	32	44	100
10.694	56	32	44	72	11.250	72	24	24	64	11.852	64	24	32	72
10.713	40	28	24	32	11.250	72	32	24	48	11.905	100	28	24	72
10.714	48	32	40	56	11.250	72	32	28	56	11.905	100	48	32	56
10.714	48	28	40	64	11.313	64	44	56	72	11.905	100	56	48	72
10.714	100	40	24	56	11.314	72	28	44	100	11.938	56	24	44	86
10.714	72	48	40	56	11.363	100	44	24	48	11.944	86	24	24	72
10.750	86	40	24	48	11.363	100	44	28	56	11.960	72	28	40	86
10.750	86	40	28	56	11.363	100	44	32	64	12.000	48	24	24	40
10.750	86	40	32	64	11.401	86	44	28	48	12.000	56	28	24	40
10.750	86	32	40	100	11.429	32	24	24	28	12.000	64	32	24	40
10.800	72	32	48	100	11.429	64	28	24	48	12.000	72	40	32	48
10.853	56	24	40	86	11.429	64	24	24	56	12.000	72	24	40	100
10.859	86	44	40	72	11.429	48	24	32	56	12.031	56	32	44	64
10.909	72	44	32	48	11.454	72	40	28	44	12.040	86	40	56	100
10.909	56	28	24	44	11.459	44	24	40	64	12.121	40	24	32	44
10.909	48	24	24	44	11.459	44	32	40	48	12.121	64	44	40	48
10.909	64	32	24	44	11.467	86	24	32	100	12.153	100	32	28	72
10.913	100	56	44	72	11.467	86	48	64	100	12.153	100	64	56	72
10.937	56	32	40	64	11.512	72	32	44	86	12.178	72	44	64	86
10.937	100	40	28	64	11.518	86	28	24	64	12.216	86	44	40	64
10.945	86	44	56	100	11.518	86	32	24	56	12.222	44	24	32	48
10.949	86	48	44	72	11.518	86	56	48	64	12.222	48	24	44	72
10.972	64	28	48	100	11.520	72	40	64	100	12.222	56	28	44	72
11.000	44	24	24	40	11.574	100	48	40	72	12.222	64	32	44	72
11.021	72	28	24	56	11.629	100	24	24	86	12.245	48	28	40	56
11.057	86	56	72	100	11.638	64	40	32	44	12.250	56	32	28	40
11.111	40	24	32	48	11.667	56	24	24	48	12.272	72	32	24	44

LEAD IN INCHES	DRIVEN GEAR ON WORM	DRIVER 1ST GEAR ON STUD	DRIVEN 2ND GEAR ON STUD	DRIVER GEAR ON SCREW	LEAD IN INCHES	DRIVEN GEAR ON WORM	DRIVER 1ST GEAR ON STUD	DRIVEN 2ND GEAR ON STUD	DRIVER GEAR ON SCREW	LEAD IN INCHES	DRIVEN GEAR ON WORM	DRIVER 1ST GEAR ON STUD	DRIVEN 2ND GEAR ON STUD	DRIVER GEAR ON SCREW
12.272	72	44	48	64	12.900	86	32	48	100	13.566	100	48	56	86
12.277	100	56	44	64	12.900	86	48	72	100	13.611	56	24	28	48
12.286	86	28	40	100	12.963	56	24	40	72	13.636	48	32	40	44
12.286	86	40	32	56	12.987	100	44	32	56	13.636	100	40	24	44
12.318	86	48	44	64	13.020	100	48	40	64	13.636	72	44	40	48
12.343	72	28	48	100	13.024	56	24	48	86	13.643	64	24	44	86
12.375	72	40	44	64	13.024	64	32	56	86	13.650	86	28	32	72
12.403	64	24	40	86	13.030	86	44	32	48	13.650	86	56	64	72
12.444	64	40	56	72	13.030	86	44	48	72	13.672	100	32	28	64
12.468	64	28	24	44	13.062	64	28	32	56	13.682	86	40	28	44
12.468	48	28	32	44	13.082	100	64	72	86	13.713	64	40	48	56
12.468	64	44	48	56	13.090	72	40	32	44	13.715	64	28	24	40
12.500	40	24	24	32	13.096	44	28	40	48	13.715	48	28	32	40
12.500	48	24	40	64	13.096	44	24	40	56	13.750	44	24	24	32
12.500	56	28	40	64	13.125	72	32	28	48	13.750	48	24	44	64
12.500	100	40	24	48	13.125	72	24	28	64	13.750	56	28	44	64
12.500	100	40	28	56	13.125	56	32	48	64	13.760	86	40	64	100
12.500	100	40	32	64	13.125	72	48	56	64	13.889	100	24	24	72
12.542	86	40	28	48	13.139	86	40	44	72	13.933	86	48	56	72
12.508	86	44	64	100	13.157	72	28	44	86	13.935	86	24	28	72
12.558	72	32	48	86	13.163	86	28	24	56	13.953	72	24	40	86
12.571	64	40	44	56	13.200	72	24	44	100	13.953	100	40	48	86
12.572	44	28	32	40	13.258	100	44	28	48	13.960	86	44	40	56
12.600	72	32	56	100	13.289	100	28	32	86	13.968	64	28	44	72
12.627	100	44	40	72	13.289	100	56	64	86	14.000	56	24	24	40
12.686	100	44	48	86	13.333	64	24	24	48	14.000	48	24	28	40
12.698	64	28	40	72	13.333	64	24	28	56	14.000	64	32	28	40
12.727	64	32	28	44	13.333	56	28	32	48	14.025	72	44	48	56
12.728	56	24	24	44	13.333	56	28	48	72	14.026	72	28	24	44
12.728	48	24	28	44	13.333	64	32	48	72	14.063	72	32	40	64
12.732	100	48	44	72	13.393	100	56	48	64	14.071	86	44	72	100
12.758	64	28	48	86	13.393	100	28	24	64	14.078	86	48	44	56
12.791	100	40	44	86	13.393	100	32	24	56	14.142	72	40	44	56
12.798	86	48	40	56	13.396	72	40	64	86	14.204	100	44	40	64
12.800	64	28	56	100	13.437	86	32	28	56	14.260	56	24	44	72
12.800	64	24	48	100	13.438	86	24	24	64	14.286	40	24	24	28
12.834	56	40	44	48	13.438	86	32	24	48	14.286	48	24	40	56
12.834	44	24	28	40	13.469	48	28	44	56	14.286	64	32	40	56
12.857	72	28	32	64	13.500	72	32	24	40	14.286	100	40	32	56
12.857	72	24	24	56	13.500	72	40	48	64	14.318	72	32	28	44
12.857	72	28	24	48	13.514	86	28	44	100	14.319	72	44	56	64
12.858	48	28	24	32	13.566	100	24	28	86	14.322	100	48	44	64

LEAD IN INCHES	GEAR ON WORM (DRIVEN)	1ST GEAR ON STUD (DRIVER)	2ND GEAR ON STUD (DRIVEN)	GEAR ON SCREW (DRIVER)	LEAD IN INCHES	GEAR ON WORM (DRIVEN)	1ST GEAR ON STUD (DRIVER)	2ND GEAR ON STUD (DRIVEN)	GEAR ON SCREW (DRIVER)	LEAD IN INCHES	GEAR ON WORM (DRIVEN)	1ST GEAR ON STUD (DRIVER)	2ND GEAR ON STUD (DRIVEN)	GEAR ON SCREW (DRIVER)
14.333	86	40	32	48	15.238	64	28	48	72	15.989	100	32	44	86
14.333	86	24	40	100	15.239	64	28	32	48	16.000	64	24	24	40
14.333	86	40	48	72	15.239	64	24	32	56	16.000	48	24	32	40
14.352	72	28	48	86	15.272	56	40	48	44	16.000	56	28	32	40
14.400	72	24	48	100	15.278	44	24	40	48	16.042	56	24	44	64
14.400	72	28	56	100	15.279	100	40	44	72	16.042	56	32	44	48
14.400	72	32	64	100	15.306	100	28	24	56	16.043	44	24	28	32
14.536	100	32	40	86	15.349	72	24	44	86	16.071	72	32	40	56
14.545	64	24	24	44	15.357	86	28	24	48	16.071	72	28	40	64
14.545	48	24	32	44	15.357	86	24	24	56	16.125	86	32	24	40
14.545	56	28	32	44	15.357	86	28	32	64	16.125	86	40	48	64
14.583	56	32	40	48	15.429	72	40	48	56	16.204	100	72	28	24
14.583	56	24	40	64	15.429	72	28	24	40	16.204	100	48	56	72
14.583	100	40	28	48	15.469	72	32	44	64	16.233	100	44	40	56
14.584	40	24	28	32	15.480	86	40	72	100	16.280	100	40	56	86
14.651	72	32	56	86	15.504	100	48	64	86	16.288	86	44	40	48
14.659	86	44	48	64	15.504	100	24	32	86	16.296	64	24	44	72
14.659	86	32	24	44	15.556	64	32	56	72	16.327	64	28	40	56
14.667	64	40	44	48	15.556	64	24	28	48	16.333	56	24	28	40
14.668	44	24	32	40	15.556	56	24	32	48	16.364	72	24	24	44
14.694	72	28	32	56	15.556	32	24	28	24	16.370	100	48	44	56
14.743	86	28	48	100	15.556	56	24	48	72	16.423	86	32	44	72
14.780	86	40	44	64	15.584	48	28	40	44	16.456	72	28	64	100
14.800	100	44	56	86	15.625	100	24	24	64	16.500	72	40	44	48
14.815	64	24	40	72	15.625	100	32	24	48	16.500	48	32	44	40
14.849	56	24	28	44	15.625	100	32	28	56	16.612	100	28	40	86
14.880	100	48	40	56	15.636	86	40	32	44	16.623	64	28	32	44
14.884	64	28	56	86	15.677	86	32	28	48	16.667	56	28	40	48
14.884	64	24	48	86	15.677	86	24	28	64	16.667	64	32	40	48
14.931	86	32	40	72	15.677	86	48	56	64	16.667	100	40	32	48
14.933	64	24	56	100	15.714	44	24	24	28	16.667	100	40	48	72
14.950	100	56	72	86	15.714	48	24	44	56	16.722	86	40	56	72
15.000	48	24	24	32	15.714	64	32	44	56	16.744	72	24	48	86
15.000	56	28	24	32	15.750	72	32	28	40	16.744	72	28	56	86
15.000	72	24	24	48	15.750	72	40	56	64	16.744	72	32	64	86
15.000	72	24	28	56	15.767	86	24	44	100	16.752	86	44	48	56
15.000	72	24	32	64	15.873	100	56	64	72	16.753	86	28	24	44
15.000	56	28	48	64	15.874	100	28	32	72	16.797	86	32	40	64
15.050	86	32	56	100	15.909	100	40	28	44	16.800	72	24	56	100
15.150	100	44	32	48	15.909	56	32	40	44	16.875	72	32	48	64
15.151	100	44	48	72	15.925	86	48	64	72	16.892	86	40	44	56
15.202	86	44	56	72	15.926	86	24	32	72	16.914	100	44	64	86

	DRIVEN	DRIVER	DRIVEN	DRIVER		DRIVEN	DRIVER	DRIVEN	DRIVER		DRIVEN	DRIVER	DRIVEN	DRIVER
LEAD IN INCHES	GEAR ON WORM	1ST GEAR ON STUD	2ND GEAR ON STUD	GEAR ON SCREW	LEAD IN INCHES	GEAR ON WORM	1ST GEAR ON STUD	2ND GEAR ON STUD	GEAR ON SCREW	LEAD IN INCHES	GEAR ON WORM	1ST GEAR ON STUD	2ND GEAR ON STUD	GEAR ON SCREW
16.969	64	44	56	48	17.918	86	32	48	72	19.091	72	24	28	44
16.970	64	24	28	44	17.959	64	28	44	56	19.096	100	32	44	72
16.970	56	24	32	44	18.000	72	24	24	40	19.111	86	40	64	72
17.045	100	32	24	44	18.181	56	28	40	44	19.136	72	28	64	86
17.046	100	44	48	64	18.181	64	32	40	44	19.197	86	32	40	56
17.062	86	28	40	72	18.181	100	40	32	44	19.197	86	28	40	64
17.101	86	44	56	64	18.182	48	24	40	44	19.200	72	24	64	100
17.102	86	32	28	44	18.229	100	32	28	48	19.250	56	32	44	40
17.141	64	32	48	56	18.229	100	24	28	64	19.285	72	32	48	56
17.143	64	28	24	32	18.229	100	48	56	64	19.285	72	28	48	64
17.144	48	24	24	28	18.273	100	28	44	86	19.286	72	28	24	32
17.144	72	28	32	48	18.285	64	28	32	40	19.350	86	32	72	100
17.144	72	24	32	56	18.333	56	28	44	48	19.380	100	24	40	86
17.144	72	48	64	56	18.333	64	32	44	48	19.394	64	24	32	44
17.188	100	40	44	64	18.367	72	28	40	56	19.444	40	24	28	24
17.200	86	32	64	100	18.428	86	28	24	40	19.444	56	24	40	48
17.200	86	28	56	100	18.428	86	40	48	56	19.444	100	40	56	72
17.200	86	24	48	100	18.476	86	32	44	64	19.480	100	28	24	44
17.275	86	56	72	64	18.519	100	24	32	72	19.480	100	44	48	56
17.361	100	32	40	72	18.519	100	48	64	72	19.531	100	32	40	64
17.364	64	24	56	86	18.605	100	40	64	86	19.535	72	24	56	86
17.373	86	44	64	72	18.663	100	64	86	72	19.545	86	24	24	44
17.442	100	32	48	86	18.667	64	24	28	40	19.590	64	28	48	56
17.442	100	48	72	86	18.667	56	24	32	40	19.635	72	40	48	44
17.454	64	40	48	44	18.667	64	40	56	48	19.642	100	40	44	56
17.500	56	24	24	32	18.700	72	44	64	56	19.643	44	28	40	32
17.500	48	24	28	32	18.700	72	28	32	44	19.656	86	28	64	100
17.500	72	24	28	48	18.750	100	32	24	40	19.687	72	32	56	64
17.500	56	24	48	64	18.750	72	24	40	64	19.710	86	40	44	48
17.550	86	28	32	56	18.750	72	32	40	48	19.840	100	28	40	72
17.677	100	44	56	72	18.750	100	40	48	64	19.886	100	44	56	64
17.679	72	32	44	56	18.770	86	28	44	72	19.887	100	32	28	44
17.679	72	28	44	64	18.812	86	32	28	40	19.908	86	24	40	72
17.778	64	24	32	48	18.812	86	40	56	64	19.934	100	28	48	86
17.778	64	24	48	72	18.858	48	28	44	40	20.00	72	24	32	48
17.778	64	28	56	72	18.939	100	44	40	48	20.00	64	24	24	32
17.858	100	24	24	56	19.029	100	44	72	86	20.00	56	24	24	28
17.858	100	28	32	64	19.048	40	24	32	28	20.07	86	24	56	100
17.858	100	28	24	48	19.048	64	24	40	56	20.09	100	56	72	64
17.917	86	24	32	64	19.048	64	28	40	48	20.16	86	48	72	64
17.917	86	24	28	56	19.090	56	32	48	44	20.16	86	32	48	64
17.918	86	24	24	48	19.090	72	44	56	48	20.20	100	44	64	72

LEAD IN INCHES	DRIVEN — GEAR ON WORM	DRIVER — 1ST GEAR ON STUD	DRIVEN — 2ND GEAR ON STUD	DRIVER — GEAR ON SCREW
20.20	72	28	44	56
20.35	100	32	56	86
20.36	64	40	56	44
20.41	100	28	32	56
20.42	56	24	28	32
20.45	72	32	40	44
20.48	86	48	64	56
20.48	86	28	48	72
20.48	86	28	32	48
20.48	86	24	32	56
20.57	72	40	64	56
20.57	72	28	32	40
20.63	72	32	44	48
20.63	72	24	44	64
20.74	64	24	56	72
20.78	64	28	40	44
20.83	100	32	48	72
20.83	100	24	32	64
20.83	100	24	28	56
20.83	100	24	.24	48
20.90	86	32	56	72
20.90	86	24	28	48
20.93	100	40	72	86
20.95	64	28	44	48
20.95	64	24	44	56
20.95	44	24	32	28
21.00	56	32	48	40
21.00	72	40	56	48
21.00	72	24	28	40
21.12	86	32	44	56
21.12	86	28	44	64
21.21	56	24	40	44
21.32	100	24	44	86
21.33	100	56	86	72
21.33	64	24	32	40
21.39	44	24	28	24
21.39	56	24	44	48
21.43	100	40	48	56
21.43	72	28	40	48
21.43	72	24	40	56
21.43	48	28	40	32
21.43	100	28	24	40
21.48	100	32	44	64
21.50	86	24	24	40
21.82	72	44	64	48
21.82	100	28	44	72
21.82	64	32	48	44
21.82	56	28	48	44
21.82	72	24	32	44
21.88	100	40	56	64
21.88	100	32	28	40
21.90	86	24	44	72
21.94	86	28	40	56
21.99	86	44	72	64
22.00	64	32	44	40
22.00	48	24	44	40
22.00	56	28	44	40
22.04	72	28	48	56
22.11	86	28	72	100
22.22	100	40	64	72
22.22	40	24	32	24
22.22	64	24	40	48
22.32	72	24	64	86
22.32	100	32	40	56
22.32	100	28	40	64
22.34	86	44	64	56
22.34	86	28	32	44
22.40	86	32	40	48
22.40	86	24	40	64
22.50	72	24	48	64
22.50	72	24	24	32
22.50	72	28	56	64
22.73	100	24	24	44
22.80	86	48	56	44
22.80	86	24	28	44
22.86	64	24	24	28
22.86	48	24	32	28
22.86	64	24	48	56
22.91	72	44	56	40
22.92	100	40	44	48
22.92	44	24	40	32
22.93	86	24	64	100
23.04	86	56	72	48
23.04	86	32	48	56
23.04	86	28	48	64
23.04	86	28	24	32
23.14	100	24	40	72
23.26	100	32	64	86
23.26	100	28	56	86
23.26	100	24	48	86
23.33	64	32	56	48
23.33	48	24	28	24
23.33	64	24	28	32
23.38	72	28	40	44
23.44	100	48	72	64
23.44	100	32	48	64
23.45	86	40	48	44
23.52	86	32	56	64
23.57	72	28	44	48
23.57	72	24	44	56
23.57	48	28	44	32
23.81	100	48	64	56
23.81	100	28	48	72
23.81	100	28	32	48
23.81	100	24	32	56
23.89	86	32	64	72
23.89	86	28	56	72
23.89	86	24	48	72
23.89	86	24	32	48
24.00	64	40	72	48
24.00	72	24	32	40
24.00	56	28	48	40
24.00	64	32	48	40
24.00	100	56	86	64
24.13	86	28	44	56
24.19	86	40	72	64
24.24	64	24	40	44
24.31	100	32	56	72
24.31	100	24	28	48
24.43	86	32	40	44
24.44	44	24	32	24
24.44	64	24 ·	44	48
24.54	72	32	48	44
24.55	100	32	44	56

LEAD IN INCHES	DRIVEN GEAR ON WORM	DRIVER 1ST GEAR ON STUD	DRIVEN 2ND GEAR ON STUD	DRIVER GEAR ON SCREW	LEAD IN INCHES	DRIVEN GEAR ON WORM	DRIVER 1ST GEAR ON STUD	DRIVEN 2ND GEAR ON STUD	DRIVER GEAR ON SCREW	LEAD IN INCHES	DRIVEN GEAR ON WORM	DRIVER 1ST GEAR ON STUD	DRIVEN 2ND GEAR ON STUD	DRIVER GEAR ON SCREW
24.55	100	28	44	64	26.52	100	24	28	44	28.57	100	56	64	40
24.57	86	40	64	56	26.58	100	28	64	86	28.57	48	28	40	24
24.57	86	28	32	40	26.67	64	28	56	48	28.57	64	32	40	28
24.64	86	24	44	64	26.67	56	24	32	28	28.57	100	28	32	40
24.64	86	32	44	48	26.67	48	24	32	24	28.64	72	44	56	32
24.75	72	32	44	40	26.79	48	24	72	56	28.65	100	32	44	48
24.88	100	72	86	48	26.79	100	32	48	56	28.65	100	24	44	64
24.93	64	28	48	44	26.79	100	28	48	64	28 67	86	40	64	48
25.00	72	24	40	48	26.79	100	28	24	32	28 67	86	24	32	40
25.00	48	24	40	32	26.88	86	28	56	64	29.09	64	24	48	44
25.00	56	28	40	32	26.88	86	24	48	64	29.09	64	28	56	44
25.00	100	24	24	40	26.88	86	24	24	32	29.17	100	40	56	48
25.08	86	24	28	40	27.00	72	32	48	40	29.17	56	24	40	32
25.09	86	40	56	48	27.13	100	24	56	86	29.17	100	24	28	40
25.13	86	44	72	56	27.15	100	44	86	72	29.22	100	56	72	44
25.14	64	28	44	40	27.22	56	24	28	24	29.32	86	48	72	44
25.45	64	44	56	32	27.27	100	40	48	44	29.32	86	32	48	44
25.45	56	24	48	44	27.27	72	24	40	44	29.34	64	24	44	40
25.46	100	24	44	72	27.30	86	28	64	72	29.39	72	28	64	56
25.51	100	28	40	56	27.34	100	32	56	64	29.56	86	32	44	40
25.57	100	64	72	44	27.36	86	40	56	44	29.76	100	28	40	48
25.60	86	28	40	48	27.43	64	28	48	40	29.76	100	24	40	56
25.60	86	24	40	56	27.50	56	32	44	28	29.86	100	40	86	72
25.67	56	24	44	40	27.50	48	24	44	32	29.86	86	24	40	48
25.71	72	24	48	56	27.50	72	24	44	48	29.90	100	28	72	86
25.71	72	56	64	32	27.64	86	40	72	56	30.00	56	28	48	32
25.72	72	24	24	28	27.78	100	32	64	72	30.00	72	32	64	48
25.80	86	24	72	100	27.78	100	28	56	72	30.00	72	28	56	48
25.97	100	44	64	56	27.78	100	24	48	72	30.23	86	32	72	64
25.97	100	28	32	44	27.78	100	24	32	48	30.30	100	48	64	44
26.04	100	32	40	48	27.87	86	24	56	72	30.30	100	24	32	44
26.04	100	24	40	64	27.92	86	28	40	44	30.48	64	24	32	28
26.06	86	44	64	48	28.00	100	64	86	48	30.54	100	44	86	64
26.06	86	24	32	44	28.00	64	32	56	40	30.56	44	24	40	24
26.16	100	32	72	86	28.00	56	24	48	40	30.61	100	28	48	56
26.18	72	40	64	44	28.05	72	28	48	44	30.71	86	24	48	56
26.19	44	24	40	28	28.06	100	28	44	56	30.71	86	32	64	56
26.25	72	32	56	48	28.13	100	40	72	64	30.72	86	24	24	28
26.25	72	24	56	64	28.15	86	28	44	48	30.86	72	28	48	40
26.25	72	24	28	32	28.15	86	24	44	56	31.01	100	24	64	86
26.33	86	28	48	56	28.29	72	28	44	40	31.11	64	24	56	48
26.52	100	44	56	48	28.41	100	32	40	44	31.11	56	24	32	24

	DRIVEN	DRIVER	DRIVEN	DRIVER		DRIVEN	DRIVER	DRIVEN	DRIVER		DRIVEN	DRIVER	DRIVEN	DRIVER
LEAD IN INCHES	GEAR ON WORM	1ST GEAR ON STUD	2ND GEAR ON STUD	GEAR ON SCREW	LEAD IN INCHES	GEAR ON WORM	1ST GEAR ON STUD	2ND GEAR ON STUD	GEAR ON SCREW	LEAD IN INCHES	GEAR ON WORM	1ST GEAR ON STUD	2ND GEAR ON STUD	GEAR ON SCREW
31.11	64	24	28	24	34.09	100	44	48	32	37.50	72	24	40	32
31.25	100	28	56	64	34.20	86	44	56	32	37.63	86	32	56	40
31.25	100	24	48	64	34.29	72	48	64	28	37.88	100	24	40	44
31.25	100	24	24	32	34.29	72	24	64	56	38.10	64	24	40	28
31.27	86	40	64	44	34.29	64	32	48	28	38.18	72	24	56	44
31.35	86	32	56	48	34.29	72	24	32	28	38.20	100	24	44	48
31.35	86	24	56	64	34.38	100	32	44	40	38.39	100	40	86	56
31.36	86	24	28	32	34.55	86	32	72	56	38.39	86	28	40	32
31.43	64	28	44	32	34.55	86	28	72	64	38.57	72	28	48	32
31.43	48	24	44	28	34.72	100	24	40	48	38.89	56	24	40	24
31.50	72	32	56	40	34.88	100	24	72	86	38.96	100	28	48	44
31.75	100	72	64	28	34.90	100	56	86	44	39.09	86	32	64	44
31.82	100	44	56	40	35.00	72	24	56	48	39.09	86	28	56	44
31.85	86	24	64	72	35.00	56	24	48	32	39.09	86	24	48	44
31.99	100	56	86	48	35.00	72	24	28	24	39.29	100	28	44	40
32.00	64	28	56	40	35.10	86	28	64	56	39.42	86	24	44	40
32.00	64	24	48	40	35.16	100	32	72	64					
32.09	56	24	44	32	35.18	86	44	72	40	39.49	86	28	72	56
32.14	100	56	72	40	35.36	72	32	44	28	39.77	100	32	56	44
32.14	72	28	40	32	35.56	64	24	32	24	40.00	72	24	64	48
32.25	86	48	72	40	35.71	100	32	64	56	40.00	64	28	56	32
32.25	86	40	48	32	35.71	100	24	48	56	40.00	64	24	48	32
32.41	100	24	56	72	35.72	100	24	24	28	40.00	56	24	48	28
32.47	100	28	40	44	35.83	86	32	64	48	40.00	72	24	32	24
32.58	86	24	40	44	35.83	86	28	56	48	40.18	100	32	72	56
32.73	72	32	64	44	36.00	72	32	64	40	40.18	100	28	72	64
32.73	72	28	56	44	36.00	72	28	56	40	40.31	86	32	72	48
32.73	72	24	48	44	36.00	72	24	48	40	40.31	86	24	72	64
32.74	100	28	44	48	36.36	100	44	64	40	40.72	100	44	86	48
32.74	100	24	44	56	36.46	100	48	56	32	40.82	100	28	64	56
32.85	86	24	44	48	36.46	100	24	56	64	40.91	100	40	72	44
33.00	72	24	44	40	36.46	100	24	28	32	40.95	86	28	64	48
33.33	100	24	32	40	36.67	48	24	44	24	40.95	86	24	64	56
33.33	100	48	64	40	36.67	64	24	44	32	40.96	86	24	32	28
33.33	64	24	40	32	36.67	56	24	44	28	41.14	72	28	64	40
33.33	56	24	40	28	36.86	86	28	48	40	41.25	72	24	44	32
33.33	48	24	40	24	37.04	100	24	64	72	41.67	100	32	64	48
33.51	86	28	48	44	37.33	100	32	86	72	41.67	100	28	56	48
33.59	100	64	86	40	37.33	64	24	56	40	41.81	86	24	56	48
33.79	86	28	44	40	37.40	72	28	64	44	41.81	86	24	28	24
33.94	64	24	56	44	37.50	100	48	72	40	41.91	64	24	44	28
34.09	100	48	72	44	37.50	100	32	48	40	41.99	100	32	86	64

LEAD IN INCHES	DRIVEN GEAR ON WORM	DRIVER 1ST GEAR ON STUD	DRIVEN 2ND GEAR ON STUD	DRIVER GEAR ON SCREW
42.00	72	24	56	40
42.23	86	28	44	32
42.66	100	28	86	72
42.78	56	24	44	24
42.86	100	28	48	40
42.86	72	24	40	28
43.00	86	32	64	40
43.00	86	28	56	40
43.00	86	24	48	40
43.64	72	24	64	44
43.75	100	32	56	40
43.98	86	32	72	44
44.44	64	24	40	24
44.64	100	28	40	32
44.68	86	28	64	44
44.79	100	40	86	48
44.79	86	24	40	32
45.00	72	28	56	32
45.00	72 ♪	24	48	32
45.45	100	32	64	44
45.45	100	24	48	44
45.46	100	28	56	44
45.61	86	24	56	44
45.72	64	24	48	28
45.84	100	24	44	40
45.92	100	28	72	56
46.07	86	28	72	48
46.07	86	24	72	56
46.07	86	28	48	32
46.67	64	24	56	32
46.67	56	24	48	24
46.88	100	32	72	48
46.88	100	24	72	64
47.15	72	24	44	28
47.62	100	28	64	48
47.62	100	24	64	56
47.62	100	24	32	28
47.78	86	24	64	48
47.78	86	24	32	24
47.99	100	32	86	56
47.99	100	28	86	64

LEAD IN INCHES	DRIVEN GEAR ON WORM	DRIVER 1ST GEAR ON STUD	DRIVEN 2ND GEAR ON STUD	DRIVER GEAR ON SCREW
48.00	72	24	64	40
48.38	86	32	72	40
48.61	100	24	56	48
48.61	100	24	28	24
48.86	100	40	86	44
48.89	64	24	44	24
49.11	100	28	44	32
49.14	86	28	64	40
49.27	86	24	44	32
49.77	100	24	86	72
50.00	100	28	56	40
50.00	100	24	48	40
50.00	72	24	40	24
50.00	100	32	64	40
50.17	86	24	56	40
50.26	86	28	72	44
51.14	100	32	72	44
51.19	86	24	40	28
51.43	72	28	64	32
51.43	72	24	48	28
51.95	100	28	64	44
52.08	100	24	40	32
52.12	86	24	64	44
52.50	72	24	56	32
53.03	100	24	56	44
53.33	64	24	56	28
53.33	64	24	48	24
53.57	100	28	72	48
53.57	100	24	72	56
53.57	86	24	72	48
53.57	100	28	48	32
53.75	86	24	48	32
53.75	86	28	56	32
54.85	100	28	86	56
55.00	72	24	44	24
55.28	86	28	72	40
55.56	100	24	32	24
55.56	100	24	64	48
55.99	100	24	86	64
55.99	100	32	86	48
56.25	100	32	72	40

LEAD IN INCHES	DRIVEN GEAR ON WORM	DRIVER 1ST GEAR ON STUD	DRIVEN 2ND GEAR ON STUD	DRIVER GEAR ON SCREW
56.31	86	24	44	28
57.14	100	28	64	40
57.30	100	24	44	32
57.33	86	24	64	40
58.33	100	24	56	40
58.44	100	28	72	44
58.64	86	24	72	44
59.53	100	24	40	28
59.72	86	24	40	24
60.00	72	24	64	32
60.00	72	24	56	28
60.00	72	24	48	24
60.61	100	24	64	44
61.08	100	32	86	44
61.43	86	28	64	32
61.43	86	24	48	28
62.22	64	24	56	24
62.50	100	24	72	48
62.50	100	28	56	32
62.50	100	24	48	32
62.71	86	24	56	32
63.99	100	28	86	48
63.99	100	24	86	56
64.29	100	28	72	40
64.50	86	24	72	40
65.48	100	24	44	28
65.70	86	24	44	24
66.67	100	24	64	40
67.19	100	32	86	40
68.18	100	24	72	44
68.57	72	24	64	28
69.11	86	28	72	32
69.44	100	24	40	24
69.80	100	28	86	44
70.00	72	24	56	24
71.43	100	28	64	32
71.43	100	24	48	28
71.67	86	24	64	32
71.67	86	24	56	28
71.67	86	24	48	24
72.92	100	24	56	32
74.65	100	24	86	48

	DRIVEN	DRIVER	DRIVEN	DRIVER		DRIVEN	DRIVER	DRIVEN	DRIVER		DRIVEN	DRIVER	DRIVEN	DRIVER
LEAD IN INCHES	GEAR ON WORM	1ST GEAR ON STUD	2ND GEAR ON STUD	GEAR ON SCREW	LEAD IN INCHES	GEAR ON WORM	1ST GEAR ON STUD	2ND GEAR ON STUD	GEAR ON SCREW	LEAD IN INCHES	GEAR ON WORM	1ST GEAR ON STUD	2ND GEAR ON STUD	GEAR ON SCREW
75.00	100	24	72	40										
76.39	100	24	44	24										
76.79	100	28	86	40										
80.00	72	24	64	24										
80.36	100	28	72	32										
80.63	86	24	72	32										
81.44	100	24	86	44										
81.90	86	24	64	28										
83.33	100	24	64	32										
83.33	100	24	56	28										
83.33	100	24	48	24										
83.61	86	24	56	24										
89.59	100	24	86	40										
92.14	86	24	72	28										
93.75	100	24	72	32										
95.24	100	24	64	28										
95.56	86	24	64	24										
95.98	100	28	86	32										
97.22	100	24	56	24										
107.14	100	24	72	28										
107.50	86	24	72	24										
111.11	100	24	64	24										
111.98	100	24	86	32										
125.00	100	24	72	24										
127.98	100	24	86	28										
149.31	100	24	86	24										

www.ingramcontent.com/pod-product-compliance
Lightning Source LLC
Chambersburg PA
CBHW031406180326
41458CB00043B/6625/J